Microbiology Research Advances

nova
science publishers
www.novapublishers.com

Microbiology Research Advances

Endophytic Fungi: Biodiversity, Antimicrobial Activity and Ecological Implications
Qiang-Sheng Wu, PhD (Editor)
Ying-Ning Zou (Editor)
Yong-Jie Xu, PhD (Editor)
2021. ISBN: 978-1-68507-354-1 (Softcover)
2021. ISBN: 978-1-68507-442-5 (eBook)

Pathogenic Bacteria: Pathogenesis, Virulence Factors and Antibacterial Treatment Strategies
Keith D. Watts (Editor)
2021. ISBN: 978-1-68507-422-7 (Softcover)
2021. ISBN: 978-1-68507-424-1 (eBook)

Biofilms: Advances in Research and Applications
Shane Rowland (Editor)
2021. ISBN: 978-1-68507-062-5 (Hardcover)
2021. ISBN: 978-1-68507-089-2 (eBook)

Antimicrobial Peptides: Functions, Mechanisms of Action and Role in Health and Disease
Paresh Chandra Ray, PhD (Editor)
2021. ISBN: 978-1-68507-005-2 (Hardcover)
2021. ISBN: 978-1-68507-082-3 (eBook)

Biopharmacological Activities of Medicinal Plants and Bioactive Compounds
Dr. Ajeet Singh, PhD (Editor)
Dr. Navneet (Editor)
2021. ISBN: 978-1-53619-959-8 (Hardcover)
2021. ISBN: 978-1-68507-056-4 (eBook)

More information about this series can be found at
https://novapublishers.com/product-category/series/microbiology-research-advances/

Iram Liaqat
Editor

Understanding Antibiofilm Activity

nova science publishers
www.novapublishers.com

DOI: https://doi.org/10.52305/PXEB9876

Library of Congress Cataloging-in-Publication Data

ISBN: 978-1-68507-927-7

Published by Nova Science Publishers, Inc. † New York

Contents

Preface

Biofilms have grown into a major issue that poses hazards to all life forms. This book reviews all strategies which can be employed to combat biofilms.

Chapter 1 demonstrates that microbial biofilms are generated when a polymeric matrix surrounds a collection of single or heterogeneous microbial organisms. Microorganisms can defend themselves by forming a biofilm in response to environmental challenges, such as antibiotic treatments. Biofilm-forming microorganisms cause a variety of illnesses and are hard to treat due to the antibiotic resistance of bacteria in biofilm mode. Inhibition of biofilm formation during the early stages or scattering of cells producing the biofilm are two basic tactics for controlling biofilms. Despite the fact that both techniques are widely employed, none is a comprehensive or long-term eradication method for established biofilms. As a result, finding and developing biofilm eradication agents as comprehensive, stand-alone biofilm treatment solutions has become a critical research field. Anti-biofilm chemicals such as herbal active substances, chelating agents, peptide antibiotics, antibiotics, and synthetic chemical compounds, as well as modifying their structures and modes of action, are being studied extensively. The development of biofilm resistance/tolerance mechanisms is discussed in this chapter, as well as the chemical techniques employed to remediate biofilms.

Chapter 2 discusses how biofilm resistance to antimicrobial drugs has become a global problem for medical researchers. Communications involving species of different genera of prokaryotes, as well as between prokaryotic and eukaryotic cells, are used to describe anti-biofilm actions. During the treatment of patients, frequently used medicines, antibiotics, and hybrid remedies have not attained and offered the anticipated effects. The necessity of discovering strong natural anti-biofilm techniques has increased as the demand for novel agents against resistant microorganisms has intensified. The purpose of this chapter is to provide a better understanding of the anti-biofilm activities of diverse agents, including natural agents, which suppress microbial biofilms. In addition, the chapter offers brief but succinct information on existing anti-biofilm agents and their involvement in biofilm formation. These new anti-biofilm compounds, which include phenols, sulphides, furanones, bromopyrroles, peptides, and trizoles, have a strong potential to demolish biofilms by bacteria and offer favorable impacts on medications in the future. Moreover, this chapter also highlights the mechanism of plant extracts, biological surfactants, peptides (bacterial/microbial origin), and other microbial enzymes to interfere with the QS system (quorum sensing system), routes of such host-pathogen interactions, extracellular polymer breakdown, sticky mechanism, and minimal concentration of pathogen inhibition has also been explored.

Chapter 3 discusses how even though antibiofilm tactics are helpful, they are still worth investigating because each strategy is limited by physiological parameters such as composition, osmolarity, and bacterial species. The current chapter intends to evaluate the comparative study of numerous existing antibiofilm techniques, including their future potentials for improvement of infection control systems, for in-depth analysis, understanding, and successful antibiofilm strategies. This chapter reveals that in context of the shortcomings of formerly employed anti-biofilm technologies, new research as an anti-biofilm strategy will gradually extend and create new frontiers for infection control and illness prevention.

Chapter 4 reveals that biofilms are diverse microbial societies dwelling in diverse environments. These passive microorganisms can be found on a variety of surfaces, such as a basic domestic pipe, medical equipment like urinary catheter, and even plant and animal tissues. Biofilms with extracellular matrix are more resistant to antibacterial treatments. Biofilms are responsible for a variety of infections, which have become increasingly difficult to cure due to rising antibiotic resistance. To combat their occurrence, a variety of antibiofilm tactics, particularly physical approaches, have been investigated. This chapter will look at a variety of methods for physically removing biofilms off hospital devices. Physical removal of biofilms from tools, elucidation, ultrasonic power, electric and magnetic fields are all illustrations of physical techniques. Likewise, the principle of anti-biofilm activity has been examined as well as the efficacy of adjuvant therapy to prevent further spread. Physical approaches yield impressive results, which could aid the development of schemes to reduce the prevalence of biofilms in the future.

Chapter 5 confers emerging biofilm eradication tendencies and tactics, as well as the treatment of biofilm-related medical problems. When distinct bacteria establish a microbial community, the outcome is a biofilm. Most biofilms have a complicated structure and contain a wide range of species. Cell signalling molecules (quorum sensing (QS)) are used by microorganisms to communicate. Biofilms may be detrimental, especially in biological systems, since they are difficult to remove and resistant to antimicrobials. As a result, biofilms pose a serious health risk all over the world. Physical, chemical, antibiotics, antiseptics, and other traditional biofilm-fighting treatments are often ineffective and can cause environmental concerns. Therefore, the scientific community is exploring and adopting a variety of different ways. Because of the powerful anti-microbial capabilities and mechanism of action, nanoparticles (NPs) are a viable developing solution for eradicating even the toughest and most complicated biofilms. Anti-biofilm agents are already being developed using bioactive chemicals derived from natural sources, mostly plants. Many anti-biofilm agents have been derived from plants and marine creatures, including halogenated furanones, 2-amminoimidazole alkaloids, and flavonoids. Antimicrobial peptides (AMPs) are potentially interesting options due to their broad-spectrum activity and non-specific method of action. Anti-biofilm techniques based on AMP, such as preventative and therapeutic treatments, are proving to be effective. The implementation of NPs in conjunction with these natural compounds appears to play a positive role in combating biofilm-based illnesses. Consequently, the application of antimicrobial/antibiofilm agents in adequate quantities can contribute to the destruction of biofilms and the treatment of biofilm-related diseases.

Chapter 6 examines how biofilm's pervasiveness makes its removal extremely challenging. Almost all biofilms, particularly those caused by highly resistant organisms including *Pseudomonas aeruginosa*, *Acinetobacter baumannii*, *Klebsiella pneumoniae*, *Staphylococcus aureus*, *Enterobacter* species, and *Enterococcus faecium*, cannot be destroyed with currently available medications. Insect-derived antimicrobial peptides are a revolutionary antibiofilm approach (AMPs). Insects generate AMPs in their hemolymph as an immunological response to bacterial infections. These peptides have the ability to bypass antibiotic resistance pathways and exhibit antibiofilm action via a variety of routes. There are several types of insect AMPs, but particularly cecropin A is significant in preventing uropathogenic *Escherichia coli* biofilms. Consequently, insect AMPs might be a potential intervention for biofilm inhibition. Biofilm development, insect immune responses, antimicrobial peptide groups from various insect sources, their structure, and method of action have also been studied in this chapter.

Chapter 7 will look at a variety of recently discovered nano-established biomaterials as well as the approaches that are being used to create antibiofilm surfaces. Nanocomposites and nanoparticles (NPs) have distinct antibacterial and anti-biofilm properties, making them perfect candidates for biofilm eradication. Because biofilms give resistance to standard biocides, their precise structure is essential to the virulent character in the infectious mode. Since the dawn of time, pathogenic bacteria have the poisonous ability to cause a variety of illnesses in living creatures, which may be fatal at times. Human infections induced by biofilms account for 80 percent of all infections. Medical equipment is responsible for over 60% of infections related to healthcare facilities, since it facilitates the direct colonisation of various bacteria in various human organs. Antibiotic treatment is often ineffective in combating biofilms due to the rapid mutation rate of bacteria, poor antimicrobial adherence, acidic nature, low oxygen availability, and enhanced enzyme activity. Implementing nano-improved lubricants might be a practical way to prevent the formation of biofilm. Metallic NPs have also been shown to suppress bacterial activities. Biofilms have been effectively combated by antimicrobial medication delivery using nanomaterials.

Chapter 8 discusses how microbial biofilms, which are generated by the aggregation of single or varied microbial species encased in a polymeric matrix, are a frequent way for bacteria to shield themselves against environmental challenges and antibacterial treatments. Due to the frequent evolution of antibiotic resistance in bacteria, biofilms producing microorganisms are responsible for causing a wide variety of infections which are harder to monitor. Inhibition of biofilm development at an early stage and then dispersion of biofilm-forming cells are two common ways for controlling biofilm. Despite the fact that both procedures are widely utilised for biofilm management, none is a direct treatment, full, or long-term eradication solution for persistent biofilms. As a result, exploring and developing biofilm eradication agents as comprehensive, stand-alone biofilm treatment method are becoming a critical research field. The structures, modes of action, and lowest biofilm suppressive ratios of numerous anti-biofilm molecules, such as herbal active substances, chelating agents, peptide antibiotics, antibiotics, and synthetic chemical compounds, have all been analyzed with respect to biofilm control. The present chapter focuses on the development of biofilm resistance mechanisms and presents a detailed assessment of biofilm remediation chemical techniques.

Acknowledgements

All praises to Almighty Allah Who created the universe and knows whatever is in it, hidden or evident, and Who bestowed upon me the intellectual ability and wisdom to search for knowledge.

I would like to thank and acknowledge the extraordinary effort by **Prof. Robert G. J. Edyvean**, Professor and Reader in Environmental Biotechnology, University of Sheffield, UK who put all efforts to timely review the book thoroughly with enormous skills, warmth and precision. His continuous support and vision combined with enormous skills enabled me to shape the book in final form.

Thanks are due to my parents (**Liaqat Ali Awan, Shaheena Nasreen**), my brothers (**Rizwan Liaqat Awan and Rehan Liaqat Awan**) for their moral support and source of stimulation during the completion of work.

This book would have not been possible without the hard work and sincere efforts of my students (**Noor Muhammad, Waiza Ansar and Faiza Bashir**). I am eternally grateful to **NOVA Publishers** for giving me an opportunity to learn and work with them.

Iram Liaqat

Chapter 1

Natural Products for Biofilm Control: Action Mechanisms and Innovative Proposals

Gustavo Ramos Salles Ferreira
and Thiago Henrique Napoleão*
Departamento de Bioquímica, Universidade Federal de Pernambuco,
Recife, Brazil

Abstract

Biofilms are microbial communities surrounded by an extracellular polymeric substance (EPS), formed from an intercellular communication mechanism called quorum sensing (QS). To form the biofilm, microorganisms attach to different types of surfaces such as tooth enamel, soil, plants, pipes, and medical devices. Biofilm formation provides protection against a variety of adverse conditions, including antimicrobial agents. Thus, biofilms have been related to persistent and chronic infections. In recent decades, new antimicrobial agents that can act to inhibit or eradicate these biofilms have been investigated. Natural products, mainly of plant and animal origin, have shown the ability to prevent and modulate the formation of biofilms by different species of microorganisms. They act by inhibiting EPS formation, suppressing adhesion and fixation on surfaces, and interfering with the QS mechanisms. In plants, these compounds (polyphenols, alkaloids, terpenes, peptides, and lectins, among others) can be extracted from different types of tissues and organs and may be in the form of mixtures (essential oils, extracts) or isolated. In animals, antimicrobial peptides and lectins found in different tissues have been reported as antibiofilm molecules. Innovative strategies for application of these natural

*Corresponding Author's Email: thiagohn86@yahoo.com.br.

In: Understanding Antibiofilm Activity
Editor: Iram Liaqat
ISBN: 978-1-68507-927-7

antibiofilm agents include production of functionalized films, nanoemulsions, nanoparticles, and nanocapsules. In addition, molecular docking and peptide engineering have been used to direct molecular modifications aiming to potentialize the bioactivities.

Keywords: plant extracts, secondary metabolites, lectins, antimicrobial peptides

Introduction

Biofilms are defined as microbial communities encased in a self-produced polymeric matrix that adheres to solid surfaces or air-liquid interfaces, whether abiotic or biotic (Miquel et al., 2016; Slobodnikova et al., 2016). This matrix, or extracellular polymeric substance (EPS), is constituted of polysaccharides, lipids, proteins and nucleic acids and is directly linked to the architecture and cohesion of biofilms (Fleming and Wiender et al., 2010; Cortes et al., 2011). Cells in biofilms have different phenotypes from planktonic cells (i.e., non-adhered ones), with a greater capacity for colonization of surfaces, in addition to greater resistance to external stress (Macfarlane and Dillon, 2007).

Biofilm formation has as its main steps: (I) adhesion and fixation of the microbial community on the surface through structures such as flagella or through molecular interactions such as Van der Walls forces and electrostatic interactions; (II) the formation of micro-colonies through a process of microbial cell multiplication through chemical signals; (III) the formation and maturation of the biofilm through the expression of genes for the formation of EPS and communications between microbial cells through a system of auto inductive signals called quorum sensing (QS); and (IV) the dispersion of the biofilm, a stage where the microbial cells carry out a rapid multiplication and consequent dispersion, returning to the mobile form (Vasudevan, 2014; Jamal et al., 2018).

According to Ead et al. (2018), biofilms are responsible for more than 78.2% of persistent chronic wound infections that occur in humans. Furthermore, a high incidence of biofilm formation in medical devices such as catheters, orthopedic implants, contact lenses and implantable electronic devices has been reported (Romling et al., 2012; Percival, 2017). Microbial adhesion and consequent biofilm formation on medical equipment and implants represent a major obstacle to the eradication of persistent infections.

This is because one of the most important characteristics of biofilms is their resistance to the host's immune system and to antimicrobial agents (Busscher et al., 2012). The biofilm lifestyle can increase the resistance of microorganisms by decreasing the penetration of the antibiotic, production of virulence factors, and EPS action in capturing antimicrobial molecules and inactivating them through the production of enzymes (Høiby et al., 2010; Slobodnikova et al., 2016). It is known that bacterial cells in biofilms can tolerate up to 1,000 times higher concentrations of antibiotics than when they are in the planktonic form, which limits the therapeutic options (Buhmann et al., 2016).

Biofilms are involved in several human diseases, such as prostatitis, caries, rhinosinusitis, otitis media, cystic fibrosis, endocarditis, osteomyelitis, lung infections, and urinary infections (Percival, 2017; Gebreyohannes et al., 2019). For example, biofilms formed by *Enterococcus* spp. are frequently observed in urinary tract infections, wound infections, gastrointestinal tract, and endocarditis (Ch'ng et al., 2019). Biofilms are often composed of two or more microbial species, with each of these species bringing and transmitting unique characteristics that are not present in their mono species biofilm counterparts (Flemming et al., 2016). Multispecies biofilms are widely found in medical devices (Kvich et al., 2020, Joshi et al., 2021). As an example, multispecies biofilms of uropathogenic bacteria such as *Proteus*, *Staphylococcus*, *Providencia* and *Ureaplasma* are often responsible for urinary tract infections associated with catheters and the consequent production of infectious stones (Norsworthy and Pearson, 2017). Due to the high resistance to various antimicrobials, biofilms are usually clinically treated with a combination of different drugs, which, in turn, are usually quite aggressive (De La Fuente-Nuñez et al., 2013). Because of this, recent studies have sought new strategies for preventing and eradicating biofilms (Howlim et al., 2015; Gugala et al., 2017). In an effort to find agents capable of preventing and eradicating biofilms, recent studies have evaluated natural products, including plant secondary metabolites (alkaloids, terpenes, polyphenols, tannins) as well as bioactive proteins and peptides of plant and animal origin (Sun et al., 2013; Song et al., 2018; Lu et al., 2019; Melander et al., 2020). In addition, the use of mixtures of different compounds has the advantage of bringing together different modes of action, thus becoming less favorable to the emergence of resistance compared to treatment with classic antibiotics (Sun et al., 2013). Studies demonstrate the ability to combine physical methods, such as low-frequency ultrasound, with antibiotics for the removal and treatment of biofilms (Carmen et al., 2005; Dong et al., 2013).

Another target is the QS system that promotes communication between bacteria as well as modulating several genes, such as those responsible for the expression of virulence factors and formation of biofilm (Uroz et al., 2009). QS inhibitors could attenuate the action of self-inducing signals and therefore have been used both alone and in association with antibiotics (Sun et al., 2013). Natural compounds have shown anti-QS action, which will be discussed further in the next sections.

Natural Products with Antibiofilm Action

Plant Extracts, Metabolites and Essential Oils

Extracts and essential oils derived from the most diverse plant tissues and organs have been shown to be effective alternatives against biofilms. Secondary metabolites are bioactive compounds produced by plants that, among other functions, act as defensive agents, protecting them against pathogens and pests present in the plant ecosystem (Song et al., 2018). Secondary metabolites such as polyphenols, flavonoids, phenolic acids, and tannins have already demonstrated antibiofilm activity (Table 1) (Slobodnikova et al., 2016). We present some reports from the literature of the antibiofilm activities of plant extracts and isolated metabolites according to the microbial species affected.

A methanol extract of *Marsdenia tinctoria* fruits demonstrates an intense ability to inhibit biofilm formation by AmpC β-lactamases-producing *Klebsiella pneumoniae* (Satish et al., 2014). Cyanidin, a secondary metabolite of the anthocyanin family, shows the ability to inhibit biofilm formation by *K. pneumoniae* through binding to the LasR protein receptor, which is a type 2 QS regulatory molecule involved in the formation of the biofilm by this bacterium (Gopu et al., 2016).

Studies have shown the ability of some plant extracts to reduce the adhesion of microbial cells to surfaces by inhibiting the production of water-insoluble glycans that directly contribute to the formation of biofilms by caries-causing microorganisms, such as *Streptococcus mutans* (Ooshima et al., 2000; Song et al., 2006). The polyphenols pinocembrin and apigenin, present in the Chilean Propolis extract, were able to significantly reduce the biofilm production by *S. mutans* in synthetic models of caries plaques, even with higher activity than chlorhexidine (Veloz et al., 2019). *Acmella oleracea* extract rich in the compound spilanthol (a fatty acid amide) shows strong antibiofilm activity against *S. mutans*, promoting an inhibition of ca. 51% after only 1-min application. The authors considered this extract an affordable coadjuvant to prevent and control dental caries (Peretti et al., 2021). A systematic review conducted by Cardoso et al. (2021) highlighted the use of *Camellia sinensis* leaf extract as an adjuvant for the treatment of oral biofilms, including dental plaque and gingivitis; however, the authors pointed out that a better understanding of the mechanisms of action are needed to reach better results in clinical treatments.

Acroptilon repens ethanolic extract, as well as its chloroform, ethyl acetate, and water fractions, shows antibiofilm activity and a decrease in the expression level of *exoA* gene (encoding exotoxin A) in *P. aeruginosa*, and are considered new sources of compounds to fight nosocomial infections caused by drug-resistant strains of this bacterium (Akhgari et al., 2021). Umbelliferon derivatives, found in plants of *Ferula* genus, show high antibiofilm activity against *P. aeruginosa*, but did not show anti-QS action as indicated by weak inhibition of QS-controlled production of the pigment violacein (Amin et al., 2020).

Methanolic extract of *Kalanchoe laxiflora* flowers exhibits antibiofilm activity against *Escherichia coli*. It was detected the presence of flavonoids, phenolic acids, gallotannins, alkaloids and fatty acid in this extract (Osman et al., 2021). *Polypodium vulgare* rhizome aqueous extract and osladin (a saponin) isolated from it has inhibitory effects on the survival and biofilm formation by *E. coli* (Gleńsk et al., 2019).

Oncocalyxone A, a benzoquinone obtained from *Auxemma oncocalyx*, gaveantibiofilm activity against methicillin-resistant *Staphylococcus aureus* (Silva et al., 2020). *Decocts from Camellia sinensis* (non-fermented leaves and buds) and *Curcuma longa* (rhizomes) showed antibiofilm activity especially against *S. aureus*, inhibiting cell motility and violacein production. *C. longa* sample also disrupted signal reception of externally supplied acyl homoserine lactone, a signaling molecule involved in bacterial QS (Tamfu et al., 2020). In the study by Dong et al. (2018), tannic acid, a phenolic compound found in several plant species, was able to inhibit biofilm formation by *S. aureus*, both in sensitive and methicillin-resistant strains (MRSA). Trentin et al. (2011) demonstrated that aqueous extracts of plants from the Brazilian semi-arid region, such as branches and fruits of *Bauhinia acuruana* and leaves of *Pityrocarpa moniliformis*, inhibited biofilm formation of *Staphylococcus epidermidis* by 77–81% at a concentration of 4 mg/mL.

The flavonoid curcumin was able to inhibit the formation of monospecies and multispecies biofilms by *Acinetobacter baumannii* and *Candida albicans*, binding to a regulatory molecule (BfmR), a response regulator in a two-component signal transduction system in this species (Raorane et al., 2019). Saponins obtained from *Sapindus mukorossi* pulps, called sapindoside A and sapindoside B, showed antibiofilm activity against *Cutibacterium acnes*. Microscopy analysis revealed that the structure of biofilm was destroyed, with the thickness remarkably reduced. Average diffusion distance of biofilm also decreased significantly in treatments with saponins and there was a down-regulation in the expression of several biofilm-associated genes, including those encoding von Willebrand factor type A domains, proteins involved in EPS biosynthesis, and triacylglycerol lipase (Wei et al., 2022).

According to the European Pharmacopeia (7th edition), essential oils are defined as odorant products that have a complex composition and can be obtained from plant raw extract, either extracted by steam of water, dry distillation, or a suitable mechanical method without heating (El Asbahani et al., 2015). Essential oils have antimicrobial activity against the most diverse pathogens, being widely used in the food industry to extend the shelf life of products and protect against contamination by pathogens (Fratianni et al., 2010; Kalemba et al., 2013). *Coriandrum sativum* essential oil caused a strong inhibition of biofilm formation by *Stenotropomonas maltophilia* (Kacaniova et al., 2020). Schillaci et al. (2008) demonstrated that the essential oil from *Boswellia papyrifera* gives antibiofilm activity against *S. epidermidis* and *S. aureus* while the essential oil of *Cymbopogon nardus* and its major compound geraniol reduced the biofilm biomass of *S. aureus* up to 100% (Pontes et al., 2018).

Essential oil of *Tetragastris catuaba* leaves inhibit biofilm formation and planktonic growth of *Listeria monocytogenes* and *Enterococcus faecalis, both in* free and nanoemulsion-encapsulated forms. Its major compound, β-caryophyllene, was also evaluated in free and encapsulated form, but showed lower activity than the oil (Silva et al., 2021). An essential oil from *Citrus changshan-huyou* was also active on biofilm of *L. monocytogenes*, being able to decrease biomass and cellular activity. The authors also demonstrated that this oil caused extensive disruption of biofilm architecture and, using KEGG Annotation, found 23 differentially expressed genes involved in QS pathways (Guo et al., 2019). Cinnamon and cardamom essential oils inhibited *E. coli* and *Bacillus subtilis* biofilms and the compound cinnamaldehyde was found to display the highest docking energy to interact with the biofilm matrix proteins RbmA and C receptors (Pourkhosravani et al., 2021).

A nanoemulsion produced from the essential oil of *Alhagi maurorum* was found to be active against biofilms of *P. aeruginosa*, *E. coli*, *S. aureus*, *K.*

pneumoniae, *A. baumannii*, and *Bacillus cereus* (Hassanhahian et al., 2020). Polyethylene-co-vinylacetate (EVA) films containing carvacrol or cinnamaldehyde reduced the biomass formed on polymeric films surface, compromise the normal biofilm development by *S. aureus* and *E. coli* (Nostro et al., 2012).

The antibiofilm effects of essential oils on fungi has also been described. The mandarin peel essential oil showed antibiofilm activity against *Aspergillus niger* as evidenced by absence of EPS, collapse of hyphae, disintegration of cytoplasmic organelles and folding of the cell membrane (Abdel-Aziz et al., 2019). The essential oil from *Boswellia rivae* was able to both inhibit the formation of biofilms and to eradicate preformed *C. albicans* biofilms (Schillaci et al., 2008). Chifiriuc et al. (2012) evaluated nanoparticles containing *Rosmarinus officinalis* essential oil and found that they strongly inhibited the adherence ability and biofilm development by *C. albicans* and *Candida tropicalis* to a catheter surface. Table 1 shows other examples of secondary metabolites with antibiofilm action in addition to those discussed in the text.

Antimicrobial Peptides and Proteins

Antimicrobial peptides (AMPs) are molecules usually formed by 15 to 30 amino acids that are part of the innate immune system of animals, plants, fungi and bacteria. They have the ability to bind to negatively charged molecules that make up the structure of the microbial membrane, inducing a large-scale failure or small defects that end up dissipating the transmembrane potential, causing cell death. They essentially act in cases of microorganisms with little or no growth rate, as in the case of those present in biofilms (Jorge et al., 2012). LLpepII, an AMP from *Linckia laevigata* (star fish), inhibited *C. albicans* biofilm formation without inhibiting the growth of the yeast cells (Seema and Shafreen, 2020).

AMPs are very interesting molecules to be explored in the search for antibiofilm agents to replace conventional antibiotics as their synthetic analogues, called peptidomimetics, are easier to produce, as well as have a distinct mode of action, being active against strains resistant to current antibiotics (Andrea et al., 2018; Von Borowski et al., 2018). A synthetic peptide derived from a mixture of two lactoferricin-derived peptides (human and bovine lactoferrampin) was effective at low concentrations in preventing biofilm formation and, in some cases, even eradicating pre-existing biofilm of the PAO1 strain of *P. aeruginosa* (Ramamourthy and Vogel, 2021).

Table 1. Other examples of secondary metabolites with antibiofilm action

Metabolite	Action	Microrganism	References
Asimicin	Anti-QS	*C. violaceum*	Tamfu et al. (2020)
Chalcone	Inhibition of oral biofilms	*S. mutans*	Wallock-Richards et al. (2015)
Coladonin	Biofilm formation inhibition	*E. coli*	Lee et al. (2014)
Fisetin	Biofilm formation inhibition	*S. aureus*	Durig et al. (2010)
Galic acid	Biofilm formation inhibition	*E. coli* *S. mutans*	Gao et al. (2015)
N-Cerotoyltryptamine	Anti-QS	*Chromobacterium violaceum*	Tamfu et al. (2020)
Reserpin	Biofilm formation inhibition	*K. pneumoniae*	Magesh et al. (2013)

A modification of the AMP EeCentrocin 1, from the sea-urchin *Echinus esculentus*, was made by truncating 17 amino acids in its longer chain, originating a derivative called EC1-17KV. The EC1-17KV affected *C. albicans* membrane fluidity, increasing membrane permeabilization and reducing hyphal formation, leading to suppression of biofilm formation (Ma et al., 2020). In another example, a series of halogenated derivatives of the antimicrobial peptide jelleine-I, from the royal jelly of *Apis mellifera*, were synthesized and evaluated for antibiofilm activity. All molecules showed antibiofilm activity against *S. aureus*, but the chlorine, bromine, and iodine derivatives were more efficient than the fluorine derivative (Jia et al., 2019).

Certain classes of proteins have shown varied and intense antibiofilm activity. Lectins, found in the most varied plant tissues, as well as in animals and microorganisms, are proteins with the capacity to reversibly bind carbohydrates or glycoconjugates and have shown an antibiofilm action. They can act through mechanisms such as interrupting the polymerization of EPS, preventing biofilm formation; affecting the viability of microbial cells; reducing the expression of genes related to QS mechanisms; and interacting with surfactants, enzymes and polysaccharides involved in biofilm production (Saha et al., 2014; Procópio et al., 2017; Coelho et al., 2018).

A study conducted by Vasconcelos et al. (2014) demonstrated that lectins extracted from plants and algae *Cratylia floribunda*, *Vatairea macrocarpa*, *Bauhinia bauhinioides*, *Bryothamnion seaforthii* and *Hypnea musciformis* inhibit the growth of cells of *S. epidermidis*, *S. aureus*, *Klebsiella oxytoca*, *P. aeruginosa*, *C. albicans* and *C. tropicalis* both in their planktonic forms and in biofilms. PgTeL, a chitin-binding lectin isolated from the pomegranate (*Punica granatum*) fruit sarcotesta, showed antifungal activity against *C. albicans* and *Candida krusei*, and also inhibited biofilm formation by *C. albicans*. Furthermore, PgTeL inhibited biofilm formation by *S. aureus*, *L. monocytogenes* and β-lactamase-producing *E. coli* isolates (Silva et al., 2018, 2019a, 2019b, 2021). Procópio et al. (2017) demonstrated that *Calliandra surinamensis* leaf pinnulae lectin (CasuL) was able to significantly inhibit biofilms formed by a strain of *S. aureus* resistant to oxacillin and by *Staphylococcus saprophyticcus*. Biofilm formation by non-resistant *S. aureus* isolate and *C. albicans* was also remarkably inhibited by *Alpinia purpurata* inflorescence lectin (Ferreira et al., 2018).

The water-soluble lectin from *Moringa oleifera* seeds (WSMoL) demonstrates the ability to penetrate and damage the cells of preformed *Serratia marcescens* biofilms and to reduce the number of cells in the biofilm of *Bacillus* sp. In addition, it was able to control biofilm formation on a glass surface, damaging *S. marcescens* cells and preventing *Bacillus* sp. adherence (Moura et al., 2017).

Recently, Subramaniyan et al. (2021) reported that N-myristoyltaurine capped silver nanoparticles functionalized with jacalin (the lectin from the *Artocarpus integrifolia seeds) displayed* excellent biofilm inhibitory activity, reducing both bio-volume and average biofilm thickness. The authors also demonstrated by microscopy inhibition of EPS synthesis and that cells were intact and live, suggesting a true anti-biofilm activity.

Concerning animal lectins, the type C lectin from *Bothrops jararacussu* venom shows the ability to break a pre-formed *S. epidermidis* biofilm in a time-dependent manner. In addition, the lectin inhibited biofilm formation of several mastitis-causing pathogens, including different strains of *S. aureus*, *Staphylococcus hyicus*, *Staphylococcus chromogenes*, *Streptococcus agalactiae* and *E. coli* (Klein et al., 2015). Lectins from shrimp *Penaeus semisulcatus* and serum of the pearl spot *Etroplus suratensis* disrupted biofilm formation by the aquatic pathogens *Vibrio parahaemolyticus* and *Aeromonas hydrophila* (Rubeena et al., 2019; Preetham et al., 2020). Lectins from the marine sponges *Aplysina fulva* and *Chondrilla caribensis* reduced the biomass of biofilm produced by *S. aureus*, *S. epidermidis*, and *E. coli* (Marques et al., 2018; Carneiro et al., 2019). Also, a lectin from the hemolymph of the blue swimmer crab *Portunus pelagicus* disrupts the biofilm architecture of Gram-negative bacteria (Jayanthi et al., 2017).

Conclusion

Natural products from plant and animal origin show the ability to prevent and modulate the formation of biofilms by bacteria and fungi. The full and detailed mechanisms still need to be understood for most of the compounds, but evidence points to disruption of EPS architecture formation, modulation of biofilm-associated gene expression and binding to QS system components. Innovative strategies for application of these natural antibiofilm agents include production of functionalized films, nanoemulsions, nanoparticles, and nanocapsules. In addition, molecular docking and peptide engineering have been used to direct molecular modifications aiming to potentialize the bioactivities.

References

Abdel-Aziz, M. M., Emam, T. M. and Elsherbiny, E. A. (2019). Effects of mandarin (*Citrus reticulata*) peel essential oil as a natural antibiofilm agent against *Aspergillus niger* in onion bulbs. *Postharvest Biology and Technology*, 110959.

Achudhan, D., Vijayakumar, S., Malaikozhundan, B., Divya, M., Jothirajan, M., Subbian, K. and Vaseeharan, B. (2020). The antibacterial, antibiofilm, antifogging and mosquitocidal activities of titanium dioxide (TiO2) nanoparticles green-synthesized using multiple plants extracts. *Journal of Environmental Chemical Engineering*, 104521.

Akhgari, Z., Nazari, R., Zargar, M. and Tanomand, A. (2021). Antibacterial and antibiofilm properties of *Acroptilon repens* (L.) Dc extract and its effect on exotoxin A gene expression of *Pseudomonas aeruginosa*. *Gene Reports*, 101357.

Amin, A., Hanif, M., Abbas, K., Ramzan, M., Rasheed, A., Zaman, A. and Pieters, L. (2020). Studies on effects of umbelliferon derivatives against periodontal bacteria; antibiofilm, inhibition of quorum sensing and molecular docking analysis. *Microbial Pathogenesis*, 104184.

Andrea, A., Molchanova, N. and Jenssen, H. (2018). Antibiofilm peptides and peptidomimetics with focus on surface immobilization. *Biomolecules*, 27.

Bakkali, F., Averbeck, S., Averbeck, D. and Idaomar, M. (2008). Biological effects of essential oils—a review. *Food and Chemical Toxicology*, 446–475.

Buhmann, M. T., Stiefel, P., Maniura-Weber, K. and Ren, Q. (2016). *In vitro* biofilm models for device-related infections. *Trends in Biotechnology*, 945-948.

Busscher, H. J. and van der Mei, H. C. (2012). How do bacteria know they are on a surface and regulate their response to an adhering state? *PLoS pathogens*, e1002440.

Carneiro, R. F., Viana, J. T., Torres, R. C. F., da Silva, L. T., Andrade, A. L., de Vasconcelos, M. A. and Sampaio, A. H. (2019). A new mucin-binding lectin from the marine sponge *Aplysina fulva* (AFL) exhibits antibiofilm effects. *Archives of Biochemistry and Biophysics*, 169-176.

Ch'ng, J. H., Chong, K. K., Lam, L. N., Wong, J. J. and Kline, K. A. (2019). Biofilm-associated infection by enterococci. *Nature Reviews Microbiology*, 82-94.

Chifiriuc, C., Grumezescu, V., Grumezescu, A. M., Saviuc, C., Lazăr, V. and Andronescu, E. (2012). Hybrid magnetite nanoparticles/*Rosmarinus officinalis* essential oil nanobiosystem with antibiofilm activity. *Nanoscale Research Letters*, 209.

Coelho, L. C. B. B., Silva, M. S., de Oliveira, W. F., de Moura, M. C., Pontual, E. V., Gomes, F. S., Paiva, P. M. G., Napoleão, T. H. and Correia, M. T. S. (2018) Lectins as antimicrobial agents. *Journal of Applied Microbiology*, 1238-1252.

Cortes, M. E., Consuegra, J. and Sinisterra, R. D. (2011) Biofilm formation, control and novel strategies for eradication. In: Mendez-Villas, A. (Ed.), Science against microbial pathogens: communicating current research and technological advances. *Formatex Research Centre, Badajoz*, 896-905.

Silva, R. C. S., de Souza Arruda, I. R., Malafaia, C. B., de Moraes, M. M., Beck, T. S., da Camara, C. A. G., Silva, N. H., Silva, M. V., Correia, M. T. S., Frizzo, C. P. & Machado, G. (2021). Synthesis, characterization and antibiofilm/ antimicrobial activity of nanoemulsions containing *Tetragastris catuaba* (Burseraceae) essential oil against disease-causing pathogens. *Journal of Drug Delivery Science and Technology*, 102795.

Silva, R. E., Ribeiro, F. D. O. S., de Carvalho, A. M. A., Daboit, T. C., Marinho-Filho, J. D. B., Matos, T. S., Pessoa, O. D. L., Leite, J. R. S. A., Araújo, A. R. and Soares, M. J. S. (2020). Antimicrobial and antibiofilm activity of the benzoquinone oncocalyxone A. *Microbial Pathogenesis*, 104513.

de la Fuente-Nuñez, C., Reffuveille, F., Fernández, L. and Hancock, R.E.W. (2013). Bacterial biofilm development as a multicelular adaptation: antibiotic resistance and new therapeutic strategies. *Current Opinion in Microbiology*, 580-589.

Dong, G., Liu, H., Yu, X., Zhang, X., Lu, H., Zhou, T. and Cao, J. (2018). Antimicrobial and anti-biofilm activity of tannic acid against *Staphylococcus aureus*. *Natural Product Research*, 2225-2228.

Cardoso, V. F. S., Roppa, R. H. A., Antunes, C., Moraes, A. N. S., Santi, L. and Konrath, E. L. (2021). Efficacy of medicinal plant extracts as dental and periodontal antibiofilm agents: A systematic review of randomized clinical trials. *Journal of Ethnopharmacology*, 114541.

Durig, A., Kouskoumvekaki, L., Vejborg, R. M. and Klemm, P. (2010). Chemoinformatics-assisted development of new anti-biofilm compounds. *Applied Microbiology and Biotechnology*, 309–317.

Ead, J. K., Snyder, R. J., Wise, J., Cuffy, C., Jafary, H. and Fischborn, K. (2018). Is PASH Syndrome a Biofilm Disease? A Case Series and Review of the Literature. *Wounds*, 216-223.

El Asbahani, A., Miladi, K., Badri, W., Sala, M., Addi, E. A., Casabianca, H., El Mousadik, A., Hartmann, D., Jilale, A., Renaud, F.N.R. and Elaissari, A (2015). Essential oils: From extraction to encapsulation. *International Journal of Pharmaceutics*, 220-243.

Flemming, H. C. and Wingender, J. (2010). The biofilm matrix. *Nature Reviews Microbiology*, 623-633.

Fratianni, F., de Martino, L., Melone, A., de Feo, V., Coppola, R. and Nazzaro, F. (2010). Preservation of chicken breast meat treated with thyme and balm essential oils. *Journal of Food Science*, M528–M535.

Gebreyohannes, G., Nyerere, A., Bii, C. and Sbhatu, D. B. (2019). Challenges of intervention, treatment, and antibiotic resistance of biofilm-forming microorganisms. *Heliyon*, e02192.

Gleńsk, M., Tichaczek-Goska, D., Środa-Pomianek, K., Włodarczyk, M., Wesolowski, C. A. and Wojnicz, D. (2019). Differing antibacterial and antibiofilm properties of *Polypodium vulgare* L. rhizome aqueous extract and one of its purified active ingredients–osladin. *Journal of Herbal Medicine*, 100261.

Gopu, V. and Shetty, P.H. (2016). Cyanidin inhibits quorum signalling pathway of a food borne opportunistic pathogen. *Journal of Food Science and Technology*, 968- 976.

Gugala, N., Lemire, J. A., Turner, R. J. (2017). The efficacy of different anti-microbial metals at preventing the formation of, and eradicating bacterial biofilms of pathogenic indicator strains. *The Journal of Antibiotics*, 775-780.

Guo, J., Gao, Z., Li, G., Fu, F., Liang, Z., Zhu, H. and Shan, Y. (2019). Antimicrobial and antibiofilm efficacy and mechanism of essential oil from *Citrus Changshan-huyou* YB chang against *Listeria monocytogenes*. *Food Control*, 256-264.

Hassanshahian, M., Saadatfar, A. and Masoumipour, F. (2020). Formulation and characterization of nanoemulsion from *Alhagi maurorum* essential oil and study of its antimicrobial, antibiofilm, and plasmid curing activity against antibiotic-resistant pathogenic bacteria. *Journal of Environmental Health Science and Engineering*, 1015-1027.

Høiby, N., Bjarnsholt, T., Givskov, M., Molin, S. and Ciofu, O. (2010). Antibiotic resistance of bacterial biofilms. *International Journal of Antimicrobial Agents*, 322–332.

Howlin, R. P., Brayford, M. J., Webb, J. S., Cooper, J. J., Aiken, S. S. and Stoodley, P. (2015). Antibiotic-loaded synthetic calcium sulfate beads for prevention of bacterial colonization and biofilm formation in periprosthetic infections. *Antimicrobial Agents and Chemotherapy*, 111-120.

Jamal, M., Ahmad, W., Andleeb, S., Jalil, F., Imran, M., Nawaz, M.A., Hussain, T., Ali, M., Rafiq, M. and Kamil, M. A., (2018). Bacterial biofilm and associated infections. *Journal of the Chinese Medical Association*, 7-11.

Jayanthi, S., Ishwarya, R., Anjugam, M., Iswarya, A., Karthikeyan, S. and Vaseeharan, B. (2017). Purification, characterization and functional analysis of the immune molecule lectin from the haemolymph of blue swimmer crab *Portunus pelagicus* and their antibiofilm properties. *Fish & Shellfish Immunology*, 227-237.

Jia, F., Zhang, Y., Wang, J., Peng, J., Zhao, P., Zhang, L., Yao, H., Ni, J. and Wang, K. (2019). The effect of halogenation on the antimicrobial activity, antibiofilm activity, cytotoxicity and proteolytic stability of the antimicrobial peptide Jelleine-I. *Peptides*, 56-66.

Jorge, P., Lourenço, A. and Pereira, M.O. (2012). New trends in peptide-based anti-biofilm strategies: a review of recent achievements and bioinformatic approaches. *Biofouling*, 1033-1061.

Kačániová, M., Galovičová, L., Ivanišová, E., Vukovic, N. L., Štefániková, J., Valková, V. and Tvrdá, E. (2020). Antioxidant, antimicrobial and antibiofilm activity of coriander (*Coriandrum sativum* L.) essential oil for its application in foods. *Foods*, 282.

Kalemba, D. and Knicka, A. (2003). Antibacterial and antifungal properties of essential oils. *Current Medicinal Chemistry*, 813-829.

Klein, R. C., Fabres-Klein, M. H., de Oliveira, L. L., Feio, R. N., Malouin, F. and Ribon, A. D. O. B. (2015). A C-type lectin from *Bothrops jararacussu* venom disrupts Staphylococcal biofilms, *PLoS ONE*, 10, e0120514.

Lee, J. H., Kim, Y. G., Cho, H. S., Ryu, S. Y., Cho, M. H. and Lee, J. (2014). Coumarins reduce biofilm formation and the virulence of *Escherichia coli* O157:H7. *Phytomedicine*, 1037–1042.

Lu, L., Hu, W., Tian, Z., Yuan, D., Yi, G., Zhou, Y. and Li, M. (2019). Developing natural products as potential anti-biofilm agents. *Chinese Medicine*, 11.

Ma, L., Ye, X., Sun, P., Xu, P., Wang, L., Liu, Z. and Zhou, C. (2020). Antimicrobial and antibiofilm activity of the EeCentrocin 1 derived peptide EC1-17KV via membrane disruption. *EBioMedicine*, 102775.

Macfarlane, S. and Dillon, J. F. (2007). Microbial biofilms in the human gastrointestinal tract. *Journal of Applied Microbiology*, 1187–1196.

Magesh, H., Kumar, A., Alam, A., Priyam, Sekar, U., Sumatran, V.N. and Vaidyanathan, R. (2013). Identification of natural compounds which inhibit biofilm formation in clinical isolates of *Klebsiella pneumoniae*. *Indian Journal of Experimental Biology*, 764-772.

de Almeida, A. S., de Oliveira Sousa, A. R., Pereira, R., Andrade, A. L., Chaves, R. P., Carneiro, R. F., de Vasconcelos, M. A., do Nascimento-Neto, L. G., Pinheiro, U. and Videira, P. A., (2018). Antibacterial activity of a new lectin isolated from the marine sponge *Chondrilla caribensis*. *International Journal of Biological Macromolecules*, 1292-1301.

Melander, R. J., Basak, A. K. and Melander, C. (2020). Natural products as inspiration for the development of bacterial antibiofilm agents. *Natural Product Reports*, 1454-1477.

Miquel, S., Lagrafeuille, R., Souweine, B. and Forestier, C. (2016). Anti-biofilm activity as a health issue. *Frontiers in Microbiology*, 592.

Moura, M. C., Trentin, D. S., Napoleão, T. H., Primon-Barros, M., Xavier, A. S., Carneiro, N. P., Paiva, P. M. G., Macedo, A. J. and Coelho, L. C. B. B. (2017). Multi-effect of the water-soluble Moringa oleifera lectin against *Serratia marcescens* and *Bacillus* sp.: antibacterial, antibiofilm, and anti-adhesive properties. *Journal of Applied Microbiology*, 861-874.

Nostro, A., Scaffaro, R., D'Arrigo, M., Botta, L., Filocamo, A., Marino, A. and Bisignano, G. (2012). Study on carvacrol and cinnamaldehyde polymeric films: mechanical properties, release kinetics and antibacterial and antibiofilm activities. *Applied Microbiology and Biotechnology*, 1029-1038.

Ooshima, T., Osaka, Y., Sasaki, H., Osawa, K., Yasuda, H., Matsumura, M., Sobue, S. and Matsumoto, M. (2000) Caries inhibitory activity of cacao bean husk extract in in-vitro and animal experiments. *Archives of Oral Biology*, 639- 645.

Osman, E. E., Mohamed, A. S., Elkhateeb, A., Gobouri, A., Abdel-Aziz, M. M. and Abdel-Hameed, E. S. S. (2021). Phytochemical investigations, antioxidant, cytotoxic, antidiabetic and antibiofilm activities of *Kalanchoe laxiflora* flowers. *European Journal of Integrative Medicine*, 102085.

Paiva, P. M. G., Gomes, F. S., Napoleão, T. H., Sá, R. A., Correia, M. T. S. and Coelho, L. C. B. B. (2010). Antimicrobial activity of secondary metabolites and lectins from plants. In: Mendez-Vilas, A. (Ed.). Current Research, Technology, and Education Topics in Applied Microbiology and Microbial Biotechnology, *Formatex Research Center, Badajoz*, 396-406.

Percival, S. L., Hill, K. E., Williams, D. W., Hooper, S. J., Thomas, D. W. and Costerton, J. W. A (2012). A review of the scientific evidence for biofilms in wounds. *Wound Repair and Regeneration*, 647-657.

Peretti, P., Rodrigues, E. T., Souza Junior, B. M., Bezerra, R. M., Fernández, E. G., Sousa, F. F. O. and Pinheiro, M. T. (2021). Spilanthol content of *Acmella oleracea* subtypes and their bactericide and antibiofilm activities against *Streptococcus mutans*. *South African Journal of Botany*, 17-24.

Pontes, E. K. U., Melo, H. M., Nogueira, J. W. A., Firmino, N. C. S., de Carvalho, M. G., Júnior, F. E. A. C. and Cavalcante, T. T. A. (2019). Antibiofilm activity of the essential oil of citronella (*Cymbopogon nardus*) and its major component, geraniol, on the bacterial biofilms of *Staphylococcus aureus*. *Food Science and Biotechnology*, 28, 633-639.

Pourkhosravani, E., Dehghan Nayeri, F. and Mohammadi Bazargani, M. (2021). Decoding antibacterial and antibiofilm properties of cinnamon and cardamom essential oils: a combined molecular docking and experimental study. *AMB Express*, 143.

Preetham, E., Lakshmi, S., Wongpanya, R., Vaseeharan, B., Arockiaraj, J. and Olsen, R. E. (2020). Antibiofilm and immunological properties of lectin purified from shrimp *Penaeus semisulcatus*. *Fish & Shellfish Immunology*, 776-782.

Procópio, T. F., Patriota, L. L. S., Moura, M. C., Silva, P. M., Oliveira, A. P. S., Carvalho, L. V. N., Lima, T. A., Soares, T., Silva, T. D., Coelho, L. C. B. B., Pitta, M. G. R., Rêgo, M. J. B. M., Figueiredo, R. C. B. Q., Paiva, P. M. G. and Napoleão, T. H. (2017). CasuL: A new lectin isolated from *Calliandra surinamensis* leaf pinnulae with cytotoxicity to cancer cells, antimicrobial activity and antibiofilm effect. *International Journal of Biological Macromolecules*, 419-429.

Ramamourthy, G. and Vogel, H. J. Antibiofilm activity of lactoferrin-derived synthetic peptides against *Pseudomonas aeruginosa* PAO1. (2021). *Biochemistry and Cell Biology*, 138-148.

Raorane, C. J., Lee, J. H., Kim, Y. G., Rajasekharan, S. K., García-Contreras, R. and Lee, J. (2019). Antibiofilm and antivirulence efficacies of flavonoids and curcumin against *Acinetobacter baumannii*. *Frontiers in Microbiology*, 990.

Romling, U. and Balsalobre, C. (2012). Biofilm infections, their resilience to therapy and innovate treatment strategies. *Journal of International Medicine*, 541-561.

Saha, R.K., Acharya, S., Jamiruddin, M., Roy, P., Islam, M.D.S., and Shovon, S.S.H. (2014). Antimicrobial effects of a crude plant lectin isolated from the stem of *Tinospora tomentosa. The Journal of Phytopharmacology*, 44-51.

Satish, K. R., Ramesh, S., Sucharitha, K. M., Vinoth, J. (2014). Antibiofilm activity of *Morinda tinctoria* fruit extracts against AmpC-lactamase positive *Klebsiella pneumoniae*. *Der Pharmacia Lettre*, 160-165.

Schillaci, D., Arizza, V., Dayton, T., Camarda, L. and Stefano, V. D. (2008). *In vitro* anti-biofilm activity of *Boswellia* spp. oleogum resin essential oils. *Letters in Applied Microbiology*, 433-438.

Shao, D., Li, J., Li, J., Tang, R., Liu, L., Shi, J., Huang, Q. and Yang, H. (2015). Inhibition of gallic acid on the growth and biofilm formation of *Escherichia coli* and *Streptococcus mutans*. *Journal of Food Science*, 1299–1305.

Siddharthan, S. and Rajamohamed, B. S. (2020). Investigation of potential antibiofilm properties of Antimicrobial Peptide (AMP) from *Linckia laevigata* against *Candida albicans*: An *in vitro* and *in vivo* study. *Process Biochemistry*, 340-347.

Silva, L. N., Zimmer, K. R., Macedo, A. J. and Trentin, D. S. (2016). Plant natural products targeting bacterial virulence factors. *Chemiacal Reviews*, 9162–9236.

Slobodníková, L., Fialová, S., Rendeková, K., Kováč, J. and Mučaji, P. (2016). Antibiofilm activity of plant polyphenols. *Molecules*, 1717.

Song, J.H., Kim, S.K., Chang, K.W., Han, S.K., Yi, H.K. and Jeon, J.G. (2006). *In vitro* inhibitory effects of *Polygonum cuspidatum* on bacterial viability and virulence factors of *Streptococcus mutans* and *Streptococcus sobrinus*. *Archives of Oral Biology*, 1131-1140.

Song, X., Xia, Y. X., He, Z. D. and Zhang, H. J. (2018). A review of natural products with anti-biofilm activity. *Current Organic Chemistry*, 789-817.

Subramaniyan, S. B., Megarajan, S., Dharshini, K. S. and Veerappan, A. (2021). *Artocarpus integrifolia* seed lectin enhances membrane damage, oxidative stress and biofilm inhibition activity of silver nanoparticles against *Staphylococcus aureus*. *Colloids and Surfaces A: Physicochemical and Engineering Aspects*, 126842.

Sun, F., Qu, F., Ling, Y., Mao, P., Xia, P., Chen, H. and Zhou, D. (2013). Biofilm-associated infections: antibiotic resistance and novel therapeutic strategies. *Future Microbiology*, 877-886.

Tamfu, A. N., Ceylan, O., Fru, G. C., Ozturk, M., Duru, M. E. and Shaheen, F. (2020). Antibiofilm, antiquorum sensing and antioxidant activity of secondary metabolites from seeds of *Annona senegalensis*, Persoon. *Microbial Pathogenesis*, 104191.

Tamfu, A. N., Ceylan, O., Kucukaydin, S. and Duru, M. E. (2020). HPLC-DAD phenolic profiles, antibiofilm, anti-quorum sensing and enzyme inhibitory potentials of *Camellia sinensis* (L.) O. Kuntze and *Curcuma longa* L. LWT, 110150.

Trentin, D.S., Giordani, R.B., Zimmer, K.R., da Silva, A.G., da Silva, M.V., Correia, M.T.S., Baumvol, I.J.R. and Macedo, A.J. (2011). Potential of medicinal plants from the Brazilian semi-arid region (Caatinga) against *Staphylococcus epidermidis* planktonic and biofilm lifestyles. *Journal of Ethnopharmacology*, 137, 327-335.

Vasudevan, R. (2014). Biofilms: microbial cities of scientific significance. *Journal of Microbiology & Experimentation*, 84-98.

Veloz, J. J., Alvear, M. and Salazar, L. A. (2019). Antimicrobial and antibiofilm activity against *Streptococcus mutans* of individual and mixtures of the main polyphenolic compounds found in Chilean propolis. *BioMed Research International*, 7602343.

Von Borowski, G.R., Gnoatto, S. C. B., Macedo, A. J. and Gillet, R. (2018). Promising antibiofilm activity of peptidomimetics. *Frontiers in Microbiology*, 2157.

Wallock-Richards, D. J., Marles-Wright, J., Clarke, D. J., Maitra, A., Dodds, M., Hanley, B. and Campopiano, D. J. (2015). Molecular basis of *Streptococcus mutans* sortase A inhibition by the flavonoid natural product trans-chalcone. *Chemical Communications*, 10483–10485.

Wei, M. P., Yu, H., Guo, Y. H., Cheng, Y. L., Xie, Y. F. and Yao, W. R. (2021). Synergistic combination of Sapindoside A and B: A novel antibiofilm agent against *Cutibacterium acnes*. *Microbiological Research*, 126912.

Wrzosek, L., Miquel, S., Noordine, M.-L., Bouet, S., Joncquel Chevalier-Curt, M. and Robert, V. (2013). *Bacteroides thetaiotaomicron* and *Faecalibacterium prausnitzii* influence the production of mucus glycans and the development of goblet cells in the colonic epithelium of a gnotobiotic model rodent. *BMCs Biology*, 61.

Chapter 2

An Understanding of Antibiofilm Activity

Nazish Mazhar Ali[1], Bushra Mazhar[1], Maza Rehman[1], Sikander Ali[2] and Iram Liaqat[1,*]

[1]Department of Zoology, Government College University, Lahore, Pakistan
[2]Institute of Industrial Biotechnology, Government College University, Lahore, Pakistan

Abstract

The resistance of biofilms against antimicrobial agents has created a global threat/challenge to medical researchers across the world. Anti-biofilm activities are identified and characterized not only by communications among species of different genera of prokaryotes but also among prokaryotic and eukaryotic cells. Commonly used drugs, antibiotics and hybrid therapies have not achieved and delivered the desired results during treatment of patients. The search for new agents against resistant microbes has increased the importance of finding of novel and potent natural anti-biofilm tools. This chapter is designed to create better understanding of anti-biofilm activities by various agents including natural agents to inhibit the microbial biofilms and also to provide concise information on existing anti-biofilm agents as well as their role against biofim forming bacteria. Modifying natural anti-biofilm agents is one way to develop candidates for clinical exploration. In the current era, commonly used methods to modify medical devices and their surfaces for prevention of biofilm formation include newly developed antibiotics and various chemicals. New anti-biofilm agents containing components like phenols, sulfides, furanones, bromopyr roles, peptides and trizoles have good capacity to disintegrate bacterial biofilms and can

*Corresponding Author's Email: iramliaq@hotmail.com.

In: Understanding Antibiofilm Activity
Editor: Iram Liaqat
ISBN: 978-1-68507-927-7

have a positive effect on medicines of the future. Infectious microorganisms and their ability to cause infections are main concerns in biological field. Infections associated with biofilms are persistent and not easy to treat because of resistance of the microbes in biofilm mode. Less efficacy of therapies and internal toxicity of commonly available antibiotics have driven scientists to explore more tools against biofilm formers. Naturally occurring compounds can have more efficacy than non-natural manufactured compounds. This chapter focuses on natural agents which have anti-biofilm capabilities. Examples of such agents are plant extracts, biological surfactants, peptides and enzymes (of bacterial/microbial origin). Their mechanisms to interfere with QS system (quorum sensing system) pathways of host-pathogen interactions, extra-cellular disruption of polymers, adhesive mechanisms and minimum concentration of inhibition against pathogens will be discussed.

Keywords: biofilm former, anti-biofilm activity, anti-biofilm agents

Introduction

Biofilms

Biofilms can be described as sessile communities that are formed by microbes and are characterized by those cells that remain attached to a surface. They contain glycocalyx secreted by microbes, used for encasing the microbial communities. Biofilm glycocalyx also serves as protective and regulatory purpose. The biofilms are permeated by channels that function as circulatory system and allow the bacteria to exchange enzymes, water, nutrients and signals. Also these help in the disposal of toxins and facilitate metabolic cooperation (Kalia et al., 2011).

Examples of Biofilm Formers

In medical microbiology, many microbess are biofilm formers including both Gram positive and Gram-negative bacteria. Most importantly, *Staphylococcus epidermis*, *S. aureus*, *Escherichia coli*, *Pseudomonas aeruginosa*, *Streptococcus viridans*, *Enterococcus faecalis*, *Proteus mirabilis* and *Klebsiella pneumoniae* are known to form biofilms on medical instruments and implants, making these resistant to destruction and leading to problematic chronic diseases.

Anti-Biofilm Activity

Biofilms show their effects in four ways: They facilitate the activity of antimicrobial drug resistance causing chronic infections, modulate the host immune response and cause contamination of medical devices. These effects make it urgent to develop better anti-biofilm agents against them. Anti-biofilm active agents cause modulation of biofilm formers and interact with the diversity of biofilm forming microbes by interfering between species of bacteria or between the host surface and bacteria. These agents include broad spectrum peptides, antibiotics and QS inhibitors that disrupt the development of biofilm forming microbes (Liaqat et al., 2008; 2009; 2010; 2012; 2014).

Anti-Biofilm Activity of Various Agents against Biofilm Formers

Anti-biofilm activity of various agents against biofilms is described below.

Copper Oxide and Iron Oxide Nanoparticles Activity against Multidrug Resistant Uropathogens

Nanoparticles of metal oxides such as copper and iron oxides show effectiveness against those biofilm forming bacteria that are resistant against many drugs (Tufail and Liaqat, 2021; Tufail et al., 2022). Both metal oxides show anti-biofilm activity against methicillin resistant *S. aureus* (MRSA) and *E. coli*. Iron oxide is more effective as anti-biofilm agent than copper oxide and shows dose dependent anti-biofilm properties. If the dose is increased, the antibacterial properties are enhanced (Liang et al., 2012). High anti-biofilm activity was exhibited by copper oxide nanoparticles against both Gram-negative and Gram-positive bacteria (Mahanty et al., 2013). Copper oxide shows more anti-biofilm activity against Gram positive bacteria as they contain more amines and carboxyl groups and show high affinity for copper on their cell surface (Das et al., 2012).

Cathelicidin Peptides Activity (Natural and Synthetic) against S. aureus

S. aureus forms biofilms in wounds and produces a chronic infection that hinders the recovery process. Therapeutic procedures that are focused on preventing biofilm formation should be promoted (Walcott et al., 2010). Antimicrobial peptides (AMPs) are known to be essential components that take part in innate immunity in humans and various other groups of higher organisms and help in strengthening our first defensive line of immunity (James et al., 2008). Bacteria have yet to develop resistance against AMPs as they contain net positive charge, have sequences of amino acids and contain small size that help in binding and disruption of microbial membranes. It was also found that they show effects in inhibiting nucleic acid, cell wall and biosynthesis of proteins in microbes (Yeaman et al., 2003). Immuno-modulatory effects are also shown by AMPs as they show chemotactic effects for leukocytes of various kinds causing them to reach site of infection (Brogdan, 2005). They cause neutralization of lipopolysaccharides by binding to them causing angiogenesis and resulting in healing of wounds (Pollard et al., 2009).

Examples of such peptides include synthetic peptides of ceragenin class and RNA-III inhibiting peptides (Lopez et al., 2010). Cathelicidins are present in the granules of neutrophils, in lungs' mucosal epithelium and in T-cells containing anti-biofilm cathelicidin peptide. NA-CATH: ATRA1-ATRA1 is a peptide that shows high anti-microbial and anti-biofilm properties. D-LL-37 is a protease resistant peptide that shows the effectiveness of isomer of L-peptide and helps in inhibiting biofilm formation. It also shows wound healing properties (Ganz et al., 1997).

Anti-Biofilm Activity of Flavonoids against S. aureus

Staphylococci species have biofilm associated proteins known as Bap that contain amyloid like structures which act as the skeleton of the biofilm matrix (Zheng et al., 2018). Flavonoids such as quercetin, scutellarein and myricetin cause inhibition of biofilm activity of these species. These polyphenols act as anti-amyloids and help in treating infections caused by biofilm formers (Sharma et al., 2019).

Anti-Biofilm Activity of Imidazolium Salts against Candida tropicalis

Imidazolium salt and C_{16}MImCl show efficient anti-biofilm activity against multi-drug resistant *Candida tropicalis* and also show anti-fungal properties even at low concentrations and help in treating candidiasis infections (Nett et al., 2007).

Anti-Biofilm Action of Coral-Linked Bacteria for M Serotypes of Streptococcus pyogenes

Biofilm formation is greatly helped by QS (quorum sensing) molecules in different pathogens that enable cell to cell communication causing overpopulation of bacterial communities (Choo et al., 2006). The formation of biofilms by various pathogens can be controlled if QS is inhibited, on living and non-living surfaces (Doyle and Rosenberg, 1990). QS Inhibitors are present in many species of corals, sponges, tunicates and bacteria. Bacteria associated with the coral *Acropora digitifera* help in QS inhibition (Fenical et al., 2007). Combinations of oils and chemicals derived from different plants (such as oils of *Rosmarinus officinalis* and Thai plant extracts) show anti-biofilm activity against *S. mutans* and *S. pyogenes* (You et al., 2007).

Anti-Biofilm Activity of Cinnamaldehyde against Acinetobacter baumannii

A. baumannii has the ability to form biofilms and is a difficult nosocomial pathogen. It has adhesion ability and high pathogenicity. Cinnamaldehyde is an important component of cinnamomum. It has anti-biofilm properties against many Gram-positive and Gram-negative bacteria (Ryu et al., 2017).

Anti-Biofilm Properties of Manuka Honey against P. aeruginosa and S. aureus

Manuka honey is a medicinal honey obtained from Manuka bush of Australia and New Zealand. It contains low pH and high sugar content (Ryu et al., 2017). Its non-peroxide contents show useful antimicrobial properties (Adams et al., 2008). It contains methylglyoxal which predominantly shows anti-biofilm activity (Lu et al., 2013). Manuka honey shows anti-biofilm properties against *P. aeruginosa* and *S. aureus*. Its activity is increased when mixed with vancomycin and gentamicin for *S. aureus* and *P. aeruginosa,* respectively (Mavric et al., 2008). Likewise, honey derived from Asian bees also shows promising anti-biofilm potential (Liaqat et al., 2022).

The Anti-Biofilm Activity of Copper (II) Complexes

A variety of strategies are now being researched in order to improve bacterial infection treatment. There has been renewed interest in the use of metals such as copper, silver, and platinum as antibacterial agents in recent years. As an example, the use of silver in wound dressings and other products is becoming more common. Metallic copper has been found to be an efficient antibacterial surface that may be used in a variety of applications (Weaver et al., 2010). It can be used as a wound dressing or as a material for typical hospital surfaces. Furthermore, copper in paints has also been found to prevent the production of biofilms. As a result, it could be used as an antifouling agent (Cooney et al., 1999). Metallic coordination complexes with aromatic ligands are also appealing molecules with antibacterial and cytotoxic actions. The ligands are scaffolded generating a three-dimensional structure that interacts with specific targets. Dwyer's group established early on that coordination complexes of Ru(II), Ni(II), Cu(II), Fe(II), or Co(II) with 1,10-phenanthroline or other ligands show antibacterial properties (Dwyer et al., 1952).

Both Gram-positive and Gram-negative bacteria are susceptible to 2, 2'-bipyridine (Dwyer et al., 1969). These complexes interact with DNA by a mechanism called intercalation (Bolhuis et al., 2014) causing a complex formation. When compared to ligands or metals alone, biological activity of complex is dramatically increased (Zoroddu et al., 1996). Furthermore, copper (II) complexes may operate as chemical nucleases and several research groups have created and investigated DNA-facilitating compounds (Leverett et al., 2013). In recent years, the antibacterial properties of several substances containing iron, ruthenium, copper, and palladium have been investigated (Sigman et al., 1993). These are all compounds that are reported to bind DNA, either by groove binding or direct binding or intercalation. Several substances were found to have antibacterial activity against Gram-negative and Gram-positive bacteria; however antibacterial activity against the later was more effective. Mostly planktonic organisms have been used to test these complexes. As a result, Majtan et al. (2012) looked into the activity of copper (II) complexes against biofilms with particular attention on a clinical isolate of MRSA capable of producing a stable colony.

Anti-Biofilm Activity of Green Synthesized Nanoparticles of Silver against Candida Species

Candida species are opportunistic fungal pathogens that are known for infecting implanted medical equipment with biofilms in clinical situations. *Candida* species develop drug resistance as a result of broad-spectrum use of azole groups and other antifungal drugs. The majority of antifungal medications have failed to treat *Candida* infections caused by biofilms. Silver nanoparticles (AgNPs) were synthesized and evaluated for their anti-biofilm activity against *Candida* spp. AgNPs strongly reduced more than 80% of *Candida* spp. biofilm formation, according to the findings. Furthermore,

AgNPs inhibited the yeast-to-hyphal transition, exopolysaccharide production, and secreted aspartyl biosynthesis (Subramanian et al., 2018).

Anti-Biofilm Activity of the Essential Oil of Eucalyptus staigeriana

Eucalyptus staigeriana oil possesses anti-biofilm properties against both Gram positive and Gram negative bacteria. *Enterococcus faecalis* is a multi-drug resistant microbe and its biofilm can be treated by the oil of dried leaves of this plant (Adams, 2005).

Anti-Biofilm Performance of Human Milk Oligosaccharides against Acinetobacter baumannii

In immunocompromised patients, *A. baumannii* is a Gram-negative opportunistic bacterial pathogen that causes severe community and nosocomial infections. The advent of multi- and pan-drug resistant strains of *A. baumannii* complicates treatment of these infections. Biofilms, help *A. baumannii* to colonise abiotic and biotic surfaces and evade antimicrobial challenges. Biofilm-inhibiting drugs have lately acquired popularity as a chemotherapeutic method for preventing or dispersing *A. baumannii* and restoring the effectiveness of standard antimicrobials. Human milk oligosaccharides (HMOs) have recently been discovered to have antibacterial and biofilm-inhibiting effects (Sabrina et al., 2021).

P. aeruginosa Anti-Biofilm Activity of Lactoferrin-Derived Synthetic Peptides

Antimicrobial peptides (AMPs) with direct antibacterial and antifungal activities are found in human and bovine lactoferrin (hLF and bLF). They were formed entirely from the protein's highly cationic N-terminal lobe and act directly on it. Iron does not bind to microbes. Surprisinglyseveral of these peptides have been found to have an anti-inflammatory effect whereas the intact lactoferrin proteins have a significantly lower direct antibacterial activity. As a result, peptides have been proposed as prospective new antimicrobial agents (Grifford et al., 2005).

Some of these lactoferrin-derived AMPs have already been evaluated for anti-biofilm capabilities. LactoferricinB (residues 17-41) is a peptide produced from the bLF protein that helps to reduce inflammation. For example, the lactoferricinB peptide (residues 17-41) derived from the bLF protein has been shown to reduce fungal biofilms associated with keratitis (Sengupta et al., 2012), and a peptide encompassing the first eleven residues of the hLF protein (hLF11) has also been shown to reduce fungal biofilms associated with keratitis. In addition, biofilm formation by pathogenic bacteria such as *P. aeruginosa* and various periodontic bacteria can be inhibited by different lactoferrin-derived peptides (Xu et al., 2010; Morici et al., 2016).

P. aeruginosa and S. aureus Biofilm Degradation by Chitosan from Shrimps and Crabs

Crustaceans are high in chitosan, a naturally occurring polysaccharide derived from chitin. Antimicrobial activity of chitosan against a wide range of bacteria has been discovered (Goy et al., 2009). Biocompatibility and non-toxicity against humans is one of chitosan's significant biological features. Studies on chitosan as an anti-biofilm agent by penetrating biofilms, particularly in medical settings were prompted by HLs devices (Sylva et al., 2014). The creation of biofilms was inhibited by disruption of the bacteria's cell membrane's connection to the host surface. The bacteria will be killed if the biofilm is destroyed and the bacteria are made more vulnerable to antibiotics (Zhang et al., 2013).

Anti-Biofilm Activity of Satureja montana Essential Oil against Avian E. coli

Essential oils from various species of the genus *Satureja*, which belongs to the broad botanical family Lamiaceae, have been shown to have exceptional antibacterial activity against a variety of Gram-positive and Gram-negative bacteria (Chouhan et al., 2017). The volatile components, oxygenated monoterpenes such as thymol and carvacrol play a major role. The quantity is thought to be a measure of antibacterial activity (Ebani et al., 2018; Serrano et al., 2011). Instability and the emission of volatile chemicals are two issues that need to be addressed (Vittanza et al., 2019). To generate better biopharmaceutical results, qualities as well as to lower doses and negative effects while preserving volatile chemicals that demonstrate the most important antibacterial activity need to be developed (Pellegrini et al., 2018).

Anti-Biofilm Function of Moringa oleifera Lam against S. aureus and P. aeruginosa

The drumstick tree, *Moringa oleifera* Lam (Moringaceae), and *Citrus sinensis* Linn (Rutaceae) fruit have the ability to treat a variety of human infections. Extracts of *M. oleifera* and *C. sinensis* act as antimicrobial metabolites against *S. aureus* and *P. aeruginosa* where formation of the biofilm matrix is efficiently prevented (Hashemi et al., 2011).

Anti-Biofilm Effect of Chlorhexidine Conjugated Gold Nanoparticles against Klebsiella pneumonia

Chlorhexidine conjugated nanoparticles of gold not only inhibit the formation of biofilms by *K. pneumonia* but also destroy the formed biofilm. This pathogen causes nosocomial infections in urinary tract by the formation of biofilms.

Anti-Biofilm Activity of Aloe vera against Candida albicans

Aloe barbadensis leaf extracts significantly decrease biofilm formation by clinical isolates of *C. albicans*. The use of *A. vera* extracts to coat urinary catheters can reduce nosocomial infections, morbidity and mortality as well as cost burden making it a viable alternative treatment for urinary tract infections (Galleh et al., 2021).

Anti-Biofilm Activity of Essential Oil of Clove against Listeria monocytogenes and Salmonella enteritidis

In the food industry sanitizers based on the essential oil of cloves have been found to help in destroying the biofilms of *S. enteriditis* and *L. monocytogenes* (Adil et al., 2014).

Anti-Biofilm Activity of Red Ginger against C. albicans

Among other ginger varieties grown in Indonesia, such as *Zingiber officinale Roscoe* (common ginger) and *Zingiber officinale* var. amarum, red ginger has the highest essential oil content. Two main substances, Sesquiterpenes and monoterpenes extracted from the essential oils (Munfarida et al., 2018) contribute as major antibiofilm agents.

These plant-derived terpenoid chemicals are well-known for their anti-inflammatory properties. They have antibacterial and anti-biofilm properties. The essential oil of rosemary and the common ginger plant (*Zingiber officinale Roscoe*) are well-known for their medicinal properties, antimicrobial and anti-biofilm activity against bacteria and fungi (particularly *C. albicans*). Red ginger essential oil exhibits antibacterial effect against a range of microorganisms (Raut et al., 2013).

Conclusion

The control of biofilm forming capability of microbes can be done by using natural anti-biofilm agents obtained from plants and animals. These agents or products can be used as effective anti-biofilm tools against notorious biofilm formers. For many of the compounds, complete and detailed understanding of mechanisms must be explored for efficient use of these tools. There is need to study the QS system and associated genes in biofilm formers. Innovative ideas and strategies are needed to explore the potential of natural anti-biofilm agents. Also, molecular docking and protein engineering should be adapted for direct modifications at molecular level.

References

Adams, C. J., Boult, C. H., Deadman, B. J., Farr, J. M., Grainger, M. N., Manley-Harris, M. and Snow, M. J. (2008). Isolation by HPLC and characterisation of the bioactive fraction of New Zealand manuka (*Leptospermum scoparium*) honey. *Carbohydrate research*, 651-9.

Adams, R. P. (2007). Identification of essential oil components by gas chromatography/mass spectrometry. Carol Stream, IL: Allured publishing corporation.

Adil, M., Singh, K., Verma, P. K. and Khan, A. U. (2014). Eugenol-induced suppression of biofilm-forming genes in *Streptococcus mutans*: An approach to inhibit biofilms. *Journal of global antimicrobial resistance*, 286-92.

Aryal, M. and Muriana, P. M. (2019). Efficacy of commercial sanitizers used in food processing facilities for inactivation of *Listeria monocytogenes*, *E. coli* O157: H7, and *Salmonella biofilms*. *Foods*, 639.

Badawy, H. T., Pasetto, P., Mouget, J. L., Pilard, J. F., Cutright, T. J. and Milsted. A. (2013). Bacterial adhesion and growth reduction by novel rubber-derived oligomers. *Biochemical and biophysical research communications*, 691-6.

Bajalan, I., Rouzbahani, R., Pirbalouti, A. G. and Maggi, F. (2017). Variation in chemical composition and antibacterial activity of the essential oil of wild populations of Phlomis olivieri. *Chemistry & biodiversity*, e1600444.

Bolhuis, A. and Aldrich-Wright, J. R. (2014). DNA as a target for antimicrobials. *Bioorganic chemistry*, 51-9.

Brogden, K. A. (2005). Antimicrobial peptides: pore formers or metabolic inhibitors in bacteria? *Nature reviews*, 238-250. 10.1038/nrmicro1098.

Burgess, C. M., Gianotti, A., Gruzdev N, Holah. J, Knøchel. S, Lehner. A, Margas. E., Esser S. S., Sela, S. and Tresse, O. (2016). The response of foodborne pathogens to osmotic and desiccation stresses in the food chain. *International Journal of Food Microbiology*, 37-53.

Cahyono, B., Satriadi, H. and Munfarida, S. (2018). Antioxidant activity and total phenolic content in Red Ginger (Zingiber officinale) based drinks. *InIOP Conference Series: Earth and Environmental Science*. 012025.

Choo, J. H., Rukayadi, Y. and Hwang, J. K. (2006). Inhibition of bacterial quorum sensing by vanilla extract. *Letters in applied microbiology*, 637-41.

Chouhan, S.; Sharma, K. and Guleria, S. (2017). Antimicrobial activity of some essential oils—present status and future perspectives. *Medicines*, 58.

Cooney, J. J. and Tang, R. J. (1999). Quantifying effects of antifouling paints on microbial biofilm formation. *Methods in enzymology*, 637-44.

Das, A., Raychaudhuri, U. and Chakraborty, R. (2012). Antimicrobial effect of edible plant extract on the growth of some foodborne bacteria including pathogens. *Nutra Foods*, 83–88.

Doyle, R. J. and Rosenberg, M. (1990). Microbial cell surface hydrophobicity. *American Society for Microbiology*.

Dwyer, F. P., Gyarfas, E. C., Rogers, W. P. and Koch, J. H. (1952). Biological activity of complex ions. *Nature*, 190-1.

Dwyer, F. P., Reid, I. K., Shulman, A., Laycock, G. M. and Dixson, S. (1969). The biological actions of 1, 10 phenanthroline and 2, 2′ bipyridine hydrochlorides, quaternary salts and metal chelates and related compounds: 1. Bacteriostatic action on selected gram positive, gram negative and acid fast bacteria. *Australian Journal of Experimental Biology and Medical Science*, 203-18.

Ebani, V. V., Najar, B., Bertelloni, F., Pistelli, L., Mancianti, F. and Nardoni, S. (2018). Chemical composition and *in vitro* antimicrobial efficacy of sixteen essential oils against *Escherichia coli* and *Aspergillus fumigatus* isolated from poultry. *Veterinary Sciences*, 62.

Fenical, W. (1992). Chemical studies of marine bacteria: developing a new resource. *Chemical Reviews*, 1673-83.

Ganz, T. and Weiss, J. (1997). Antimicrobial peptides of phagocytes and epithelia. *Seminars in hematology*, 343-354.

Gifford, J. L., Hunter, H. N. and Vogel, H. J. (2005). Lactoferricin. *Cellular and molecular life sciences*, 2588-98.

Goy, R. C., Britto, D. D. and Assis, O. B. (2009). A review of the antimicrobial activity of chitosan. *Polímeros*, 241-7.

Hashemi, S. R. and Davoodi, H. (2011). Herbal plants and their derivatives as growth and health promoters in animal nutrition. *Veterinary research communications*, 169-80.

Kalia, V. C., Raju, S. C. and Purohit, H. J. (2011). Genomic analysis reveals versatile organisms for quorum quenching enzymes: acyl-homoserine lactone-acylase and lactonase. Open Microbiology Journal, 1-13. doi: 10.2174/1874285801105010001.

Kalia, V. C., Wood, T. K. and Kumar, P. (2013). Evolution of resistance to quorum-sensing inhibitors. *Microbiology Ecology*.

Leverett, P., Hibbs, D. E., Yang, Q., Bulanadi, J. C., Wu, M. J. and Aldrich-Wright, J. R. (2013). The antimicrobial properties of some copper (II) and platinum (II) 1, 10-phenanthroline complexes. *Dalton transactions*, 3196-209.

Liang, X., Sun, M., Li, L., Qiao, R., Chen, K., Xiao, Q. and Xu, F. (2012). Preparation and antibacterial activities of polyaniline/$Cu_{0.05}Zn_{0.95}O$ nanocomposites. *Dalton Transactions*, 2804–2811.

Liaqat, I., Arshad, N., Arshad, M., Mirza, S. A., Ali, N. M. and Shoukat, A. (2017). Antimicrobial activity of some medicinal plants extracts against food industry isolates. *Pakistan Journal of Zoology*, 565-572.

Liaqat, I., Bachmann, R. T., Sabri, A. N. and Edyvean, R. G. (2010). Isolate-specific effects of patulin, penicillic Acid and EDTA on biofilm formation and growth of dental unit water line biofilm isolates. *Current microbiology*, 148-156.

Liaqat, I., Bachmann, R. T., Sabri, A. N., Edyvean, R. G. and Biggs, C. A. (2008). Investigating the effect of patulin, penicillic acid and EDTA on biofilm formation of isolates from dental unit water lines. *Applied microbiology and biotechnology*, 349-358.

Liaqat, I., Bachmann, R. T. and Edyvean, R. G. (2014). Type 2 quorum sensing monitoring, inhibition and biofilm formation in marine microrganisms. *Current microbiology*, 342-351.

Liaqat, I., Gulab, B., Durrani A. I., Zafar, U., Rubab, S., Ali, N. M., Ulfat, M., Aftab, N. (2021). Honey potential as antibiofilm, antiquorum sensing and dispersal agent against multispecies bacterial biofilm. *Journal of Oleo Science*, 425-434.

Liaqat, I., Liaqat, M., Tahir, H. M., Ali, N. M., Arshad, M. and Arshad, N. (2019). Motility effects biofilm formation in *Pseudomonas aeruginosa* and *Enterobacter cloacae*. *Pakistan Journal of Pharmaceutical Sciences*, 927-932.

Liaqat, I., Sumbal, F. and Sabri, A. N. (2009). Tetracycline and chloramphenicol efficiency against selected biofilm forming bacteria. *Current microbiology*, 212-220.

Liebano, R. E., Abla, L. E. and Ferreira, L. M. (2008). Effect of low-frequency transcutaneous electrical nerve stimulation (TENS) on the viability of ischemic skin flaps in the rat: An amplitude study. *Wound Repair and Regeneration*, 65-9.

Lopez-Leban, F., Kiran, M. D., Wolcott, R. and Balaban, N. (2010). Molecular mechanisms of RIP, an effective inhibitor of chronic infections. *The International Journal of Artificial Organs*, 582-589.

Lu, J., Carter, D. A., Turnbull, L., Rosendale, D., Hedderley, D., Stephens, J., Gannabathula, S., Steinhorn, G., Schlothauer, R. C., Whitchurch, C. B. and Harry, E. J. (2013). The effect of New Zealand kanuka, manuka and clover honeys on bacterial growth dynamics and cellular morphology varies according to the species. *PloS one*, e55898.

Mahanty, A., Mishra, S., Bosu, R., Maurya, U. K., Netam, S. P. and Sarkar, B. (2013). Phytoextracts-synthesized silver nanoparticles inhibit bacterial fish pathogen *Aeromonas hydrophila*. *Indian Journal of Microbiology*, 438–446.

Majtan, J., Klaudiny, J., Bohova, J., Kohutova, L., Dzurova, M., Sediva, M., Bartosova, M. and Majtan, V. (2012). Methylglyoxal-induced modifications of significant honeybee proteinous components in manuka honey: possible therapeutic implications. *Fitoterapia*, 671-7.

Mavric, E., Wittmann, S., Barth, G and Henle, T. (2008). Identification and quantification of methylglyoxal as the dominant antibacterial constituent of Manuka (Leptospermum scoparium) honeys from New Zealand. *Molecular nutrition & food research*, 483-9.

Morici, P., Fais, R., Rizzato, C., Tavanti, A. and Lupetti A. (2016). Inhibition of Candida albicans biofilm formation by the synthetic lactoferricin derived peptide hLF1-11. *Plos one*, e0167470.

Nett, J., Lincoln, L., Marchillo, K., Massey, R., Holoyda, K., Hoff, B., VanHandel, M. and Andes, D. (2007). Putative role of β-1, 3 glucans in Candida albicans biofilm resistance. *Antimicrobial agents and chemotherapy*, 510-20.

Pellegrini, M., Ricci, A., Serio, A., Chaves-López, C., Mazzarrino, G., D'Amato, S., Lo Sterzo, C. and Paparella, A. (2018). Characterization of essential oils obtained from abruzzo autochthonous plants: Antioxidant and antimicrobial activities assessment for food application. *Foods*, 19.

Pollard, J., Wright, J., Feng, Y., Geng, D., Genberg, C and Savage, P. B. (2009). Activities of Ceragenin CSA-13 Against Established Biofilms in an *In Vitro* Model of Catheter Decolonization. *Anti-Infective Agents in Medicinal Chemistry*, 290-294.

Raut, J. S., Shinde. R. B., Chauhan, N. M. and Mohan Karuppayil S. (2013). Terpenoids of plant origin inhibit morphogenesis, adhesion, and biofilm formation by *Candida albicans*. *Biofouling*, 87-96.

Ryu, S. Y., Baek, W. K. and Kim, H. A. (2017). Association of biofilm production with colonization among clinical isolates of *Acinetobacter baumannii*. *Korean Journal of International Medicine*, 345–51

Schmidt, M. G., Attaway, H. H., Sharpe, P. A., John, Jr. J., Sepkowitz, K. A., Morgan, A., Fairey, S. E., Singh, S, Steed, L. L., Cantey, J. R. and Freeman, K. D. (2012). Sustained reduction of microbial burden on common hospital surfaces through introduction of copper. *Journal of clinical microbiology*, 2217-23.

Sengupta, J., Saha, S., Khetan, A., Sarkar, S. K. and Mandal, S. M. (2012). Effects of lactoferricin B against keratitis-associated fungal biofilms. *Journal of Infection and Chemotherapy*, 698-703.

Serrano, C., Matos, O., Teixeira, B., Ramos, C., Neng, N., Nogueira, J., Nunes, M. L. and Marques, A. (2011). Antioxidant and antimicrobial activity of *Satureja montana* L. extracts. *Journal of science, food and agriculture*, 1554–1560.

Sharma, D., Misba, L. and Khan, A. U. (2019). Antibiotics versus biofilm: an emerging battleground in microbial communities. *Antimicrobial Resistance & Infection Control*, 1-0.

Sigman, D. S., Mazumder, A. and Perrin, D. M. (1993). Chemical nucleases. *Chemical Reviews*, 2295-316.

Silva-Dias, A., Palmeira-de-Oliveira, A., Miranda, I. M., Branco, J., Cobrado, L., Monteiro-Soares, M., Queiroz, J. A., Pina-Vaz, C. and Rodrigues, A. G. (2014). Anti-biofilm activity of low-molecular weight chitosan hydrogel against Candida species. *Medical microbiology and immunology*, 25-33.

Vitanza, L., Maccelli, A., Marazzato, M., Scazzocchio, F., Comanducci, A., Fornarini, S., Crestoni, M. E., Filippi, A., Fraschetti, C. and Rinaldi, F. (2019). *Satureja montana* L. essential oil and its antimicrobial activity alone or in combination with gentamicin. Microbial Pathogenesis, 323–331.

Weaver, L., Noyce, J. O., Michels, H. T. and Keevil, C. W. (2010). Potential action of copper surfaces on meticillin-resistant *Staphylococcus aureus*. *Journal of applied microbiology*, 2200-5.

Wolcott, R. D. and Rhoads, D. D., Bennett, M. E., Wolcott, B. M., Gogokhia, L., Costerton, J. W. and Dowd, S. E. (2010). Chronic wounds and the medical biofilm paradigm. *Journal of wound care*, 45-53.

Xu, G. Xiong, W., Hu, Q., Zuo, P., Shao, B., Lan, F., Lu, X., Xu, Y. and Xiong, S. (2010). Lactoferrin-derived peptides and Lactoferricin chimera inhibit virulence factor production and biofilm formation in *Pseudomonas aeruginosa. Journal of applied microbiology*, 1311-8.

Yeaman, M. R. and Yount, N. Y. (2003). Mechanisms of antimicrobial peptide action and resistance. *Pharmacological reviews*, 27-55.

You, J., Xue, X., Cao, L., Lu, X., Wang, J., Zhang, L. and Zhou, S. (2007). Inhibition of Vibrio biofilm formation by a marine *actinomycete* strain A66. *Applied microbiology and biotechnology*, 1137-44.

Zhang, A., Mu, H., Zhang, W., Cui, G., Zhu, J. and Duan, J. (2013). Chitosan coupling makes microbial biofilms susceptible to antibiotics. *Scientific reports*, 1-7.

Zheng, Y., He, L., Asiamah, T. K. and Otto, M. (2018). Colonization of medical devices by staphylococci. *Environmental microbiology*, 3141-53.

Zoroddu, M. A., Zanetti, S, Pogni, R. and Basosi, R. (1996). An electron spin resonance study and antimicrobial activity of copper (II)-phenanthroline complexes. *Journal of inorganic biochemistry*, 291-300.

Chapter 3

Perspectives on Current Information and Future Potential for Antibiofilm Approaches in Health Care Systems

Aisha Waheed Qurashi[1,*], Sumaira Mazhar[1], Beenish Sarfaraz[1], Nauman Aftab[2] and Iram Liaqat[3]
[1]Department of Biology, Lahore Garrison University, Lahore, Pakistan
[2]Institute of Industrial Biotechnology, Government College University, Lahore, Pakistan
[3]Department of Zoology, Government College University, Lahore, Pakistan

Abstract

The highly competitive, dominant and complex lifestyle of pathogenic bacterial communities in the form of biofilms results in strong infections in health care systems. These biofilm infections may be associated with devices or with living tissues. With increasing antibiotic resistance in bacterial pathogens, there is an increasing demand for controlling and inhibiting biofilm formation. There are several existing effective antibiofilm strategies (utilizing natural, synthetic, organic and inorganic compounds) that remove the biofilms directly or disrupt the process of biofilm development. There are also different types of materials having antibiofilm properties. Moreover, the surface properties of materials to be used in medical devices can be modified with different compounds (natural or synthetic drugs or compounds) for short- or long-term activity. Antibiofilm activity interferes with the bacterial cell-to-cell or cell-to-surface adherence and inhibits biofilm formation effectively. Each antibiofilm strategy is delimited by various physiological factors

[*]Corresponding Author's Email: iramliaq@hotmail.com.

In: Understanding Antibiofilm Activity
Editor: Iram Liaqat
ISBN: 978-1-68507-927-7

like composition, osmolarity, bacterial species, etc. This chapter aims to review existing antibiofilm strategies, including their future potentials for improvement in infection control systems. In perspective of previously used anti-biofilm technologies and their limitations, development of novel research will progressively expand and open new horizons for infection control and disease prevention.

Keywords: antibiofilm strategies, biofilm, disease prevention, infection control

Introduction

Bacterial biofilms are one of the most problematic issues in health care systems and research efforts are focused on strategies to control biofilm infections. In public health systems, biofilms are important due to infectious diseases related to medical implants or devices. The development of a biofilm may prove significant on biotic or abiotic surfaces in environmental settings (Qurashi et al., 2012a).

This chapter focuses on strategies developed in health care systems to control biofilm infections. These infections are associated with many gram-positive, gram-negative and fungal species. Commonly reported are *Staphylococcus aureus*, *Staphylococcus epidermidis*, *Enterococcus faecalis*, *Escherichia coli, Klebsiella pneumoniae, Proteus mirabilis, Pseudomonas aeruginosa*, yeasts, and *Candida* species. Biofilms can be responsible for chronic infections. These include pulmonary infections in cystic fibrosis, surgical wound infections in diabetic patients and infections associated with medical devices like catheters, contact lenses and accessories, ventilators, dental units and waterlines (Liaqat and Sabri, 2008; Dewasthale et al., 2018). Development of biofilms of pathogenic bacterial cells results in structures that are difficult to eradicate by using standard antibiotic treatments. For example, biofilms of *S. aureus* and *S. epidermidis* are frequently isolated from cardiovascular devices and catheters (Chen et al., 2013a; Chen et al., 2013b; Subhadra et al., 2018) and *S. aureus* on contact lenses causing conjunctivitis (A Sharqi et al., 2017).

Around 30-40% of hospital acquired blood infections are nosocomial bloodstream infections caused by *S. epidermidis*, and majorly of these are through intravascular catheter infection (ICI) (Josse et al., 2019). In dental caries and periodontitis, biofilms of *Porphyromonas gingivalis* and

Streptococci spp. are the most potent biofilm formers (Eick, 2021). In dental caries the development of acidic environment by sugar metabolizing bacteria results in enamel and dentine demineralization. The complex structure so formed is so resistant that it becomes hard to eradicate (Eick, 2021).

Another important biofilm forming pathogen are *Acinetobacter* spp. that can cause severe infections in the urinary tract and soft-tissues (Jain et al., 2004), on medical implants, heart devices and joints (Subhadra et al., 2018).

It is important to note that development of biofilm is an important feature in making these infections difficult to treat with conventional antimicrobial therapies. The loosening of prosthesis mediated by biofilm development has increased with the passage of time. Species of *Staphylococcus (S. aureus*, *S.epidermidis*, and *S.lugdunensis* are commonly involved in major chronic prosthetic joint infection (PJI) Staphylococcal infections are challenging because they overcome the host immune responses (Josse et al., 2019). According to some reports (Josse et al., 2019; Rogers et al., 2009) *S. epidermidis* is a strong biofilm former causing 15-40% of surgical, prosthetic valve endocarditis (PVE) infections and up to 20% of cardiac pacemaker problems.. Development of drug resistance in biofilms may be mediated by plasmids (Pachori et al., 2019) or the development of thick exopolysaccharides that render the biofilm resistant to penetration (Subhadra et al., 2018). The development of methicillin resistance in *S. epidermidis* is a common example reported in 70% of hospitals (Halaji et al., 2017). There is thus a high demand for new antibiofilm strategies to cope with biofilms and their related infections. These strategies involve either inhibition of initial surface attachment, hindering biofilm development and maturity or disrupting biofilm communication (quorum sensing (QS)) systems (Subhadra et al., 2018). The eradication of infections becomes difficult with standard antibiotics therefore, different approaches must be used (Eick, 2021).

Biofilm Structure

Development of microbial biofilms is a natural process but variable within species (Rafi et al., 2019). Bacteria in a biofilm adopt a co-dependent lifestyle within exopolysaccharide sheath where they coordinate in complex and harmonized way and can resist unfavorable conditions like pH, salinity and osmotic stress (Liaqat et al., 2008; Qurashi et al., 2012b). The shift in lifestyle from planktonic to sessile is the first stage in the development of a biofilm. Development of biofilm starts with the attachment of bacterial cells to a surface and then results in bacterial multiplication e.g., *S. aureus* and *S. epidermidis* on medical devices (Schilcher et al., 2020).. The newly produced bacterial cells are enclosed in a sheath of extracellular matrix, protein and DNA. The surface physicochemical properties of materials, host proteins or inhibitors, can all influence biofilm formation (Lee et al., 2021).

The composition of a biofilm is mainly bacterial cells and associated exopolysaccharides (EPS). The EPS constitutes half or more of the total organic carbon and has variable chemical and physical properties (Xiao et al., 2016). The EPS is also different in different growth phases of the biofilm and contain metal ions, divalent cations, and other macromolecules from the surrounding environment. EPS may be either nonionic or ionic especially in gram-negative bacteria. The ionic property is either from different uronic acid species like D-glucuronic, D-galacturonic, and mannuronic acids or due to the presence of ketal-linked pyruvates (Sutherland, 2001) and offers great binding ability during biofilm development due to association of divalent cations. However, in gram positive bacteria, only cations contribute towards EPS composition, e.g., in *E. coli* (Evans, 2000). It is important to recognize that the EPS incorporates a considerable amount of water by hydrogen binding but may be hydrophobic or hydrophilic depending on the presence of 1, 3- or 1,4-β-linked hexose residues that makes the EPS less soluble and more rigid. The presence of organic components like proteins, DNA, lipids, humic substances, etc. in the EPS has also been reported (Nielsen et al., 1997). Species composition also influences the stability of the biofilm. For example a pure culture biofilm is generally reported as less viscous than a biofilm containing two or more species (Hussain et al., 1993).

During biofilm development, cell attachment to surfaces, a critical stage for biofilm development as well as control, happens in first 48 hours (Bertsch et al., 2021). Several factors play a decisive role in cell attachment such as exopolysaccharides, bacterial Pilli and fimbria (Javanshir et al., 2021; Nwoko et al., 2021). So control, or physical and chemical modification of surfaces can be significant during biofilm development (Ghavamian et al., 2021; Zare et al., 2021). Therefore, many trials of surface modifications or development of new surface materials are in progress. These modifications can be by using organic and inorganic nanomaterials (Mare et al., 2021); metallic oxides (Lahiri et al., 2021) or utilization of water soluble inhibiting macromolecules (Silva et al., 2021). Cell signaling plays a decisive role during biofilm formation as reported for *S. mutans by* Li et al. (2001). Another example is how the fimbria involved in cell-to-cell adhesion of many bacterial species can be altered by gene expression (*fimA*) of other bacterial species. For example,

S. cristatus has been reported to modulate the *fimA* gene expression of *P. gingivalis* by inhibiting its biofilm formation.

Biofilm Interfering Approaches

With the tendency of persistence of biofilms, eradication and control of infectious diseases becomes a great challenge. A number of strategies have been reported for the control of biofilm all with some merits and limitations (Liaqat et al., 2009a; 2009b; 2010; 2014).

Antibiotic Therapy

In many clinical studies eradication of biofilm has been found to be difficult. The only option to reduce biofilm infections is the timely replacement of devices introduced into the body, thus removing the carrier of inoculum. However, in many cases this is not feasible due to expense, risk and pain or due to differently required timings for removal as in Urinary Tract Infections (UTIs), so alternative therapies, mostly combinations of antibiotics is the only option for clinicians to suppress biofilms due to the strong resistance of biofilms to currently available antibiotics (Wu et al., 2015).

Though eradication of biofilms by antibiotics seems promising, development of resistance in bacteria, and resistance due to the biofilm structure as well as optimizing conditions for effective antibiotic therapy becomes challenging (Liaqat et al., 2009). For example, inflammation is mostly associated with bacterial infections that results in high metabolism with increased demand of oxygen at a metabolic level. Failure of sufficient oxygen provision will result in high acidosis with altered pH values that influence the efficacy of antibiotics (Wu et al., 2015). Utilisation of high dose antibiotic therapies in the presence or absence of anti-coagulants or 70 % ethanol or hydrochloric acid (2 $molL^{-1}$ HCl) therapy has been reported (Fernández-Hidalgo et al., 2014). Antibiotics in this case are reported as vancomycin (1 $mgmL^{-1}$) for Gram-positive bacteria and gentamicin (2 mgL^{-1}) for Gram-negative bacteria (Tan et al., 2014). Although antibiotic therapies are significant for biofilm eradication the stage of biofilm development will influence the treatment efficacy. For example, it is much more difficult to eradicate a mature biofilm as compared to a young biofilm, so timely use of antibiotics is preferred for quick control of infectious diseases (Hengzhuang et al., 2011; Høiby et al., 2011).

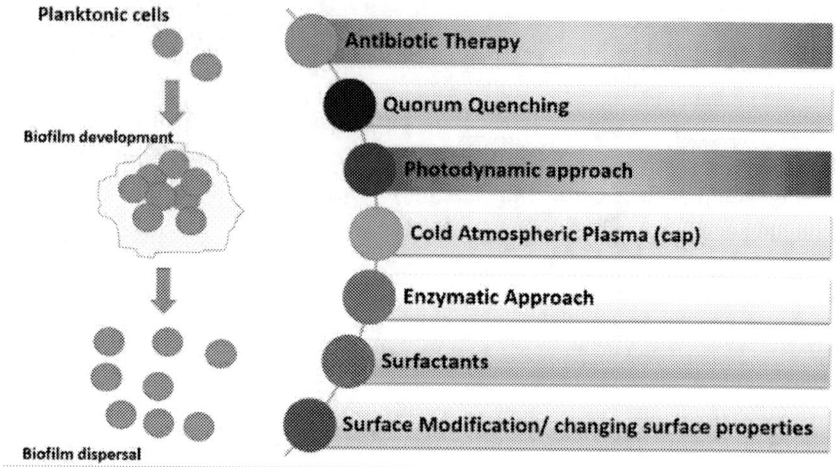

Figure 1. Diagrammatic summary of different techniques used as biofilm interfering strategies in clinical settings.

Quorum Quenching

Although conventional therapeutic measures in terms of biofilms are on their way, certain other approaches target biofilm development as well. For example, quorum sensing (QS) contributes significantly toward biofilm maturity and development by the use of signal molecules called *N*-acyl-homoserine lactones (AHL) in Gram-negative or auto-inducing peptides (AIP) in Gram-positive bacterial pathogens ((Bhatt, 2018; Srinivasan et al., 2021). The AHL synthase gene (*luxI*) is involved in AHL transcription. When produced at high concentrations, the receptor proteins (LuxR family) senses the presence of AHL molecules resulting in attachment of a signal molecule to the receptor protein resulting in the activation of the LuxR-AHL complex. This results in transcriptional control of QS mediated biofilm formation in several bacterial pathogens (Rasmussen et al., 2006).

Consequently, targeting of QS (known as quorum quenching) is a promising therapeutic approach where down-regulation or silencing of the QS system will control biofilms of Gram-negative bacterial pathogens. Utilisation of enzymes for AHL degradation is also mediated by AHL lactonases decarboxylases, acylase, and oxidase and oxidoreductase enzymes. Different bacterial species have been reported to produce these enzymes and inhibit biofilm formation in other bacterial species. AiiA 24B1 is first and well characterised AHL lactonase produced by *aiiA* gene in *Bacillus* spp. (Dong et al., 2000).

Similarly, acylase enzymes have been reported from different bacterial species with their biofilm inhibitory potential (Christensen et al., 2013; Paul et al., 2009). Interfering analogue molecules that bind themselves to the receptor proteins as QS inhibitors have also been reported to be effective for control of biofilm infections of *S. epidermidis* (Choi et al., 2012; Hume et al., 2004).

Many such molecules have been reported to be from fungal sources e.g., patulin and penicillic acid from different *Penicillium* strains (Afonso et al., 2021) and phytols (Alexpandi et al., 2021). The biofilm of *Serratia marcescens* and other pathogens can be inhibited by phytol, patulin and penicillic acid (Alexpandi et al., 2021). Advancements in research of potent QS inhibitors seems promising as adjuvant therapy to control biofilm mediated biofilm infections in health care systems (Liaqat et al., 2008; 2010; 2014).

Photodynamic Approach

Another recent technique adjuncted with traditional antibiotic therapy is to control the biofilm in health care systems using a photodynamic approach, known as photodynamic therapy (PDT) using oxygen, light, and photosensitizers. The main principle of PDT is to use photosensitizing molecules that can absorb specific light intensity and bind with target cellular components like lipids, proteins, and nucleic acids. After binding, the use of light of a suitable wavelength (700-800 nm) creates reactive oxygen radicals which further results in generation of peroxide and hydroxyl radicals and superoxide anions that becomes lethal to the cells (Darmani et al., 2018; Gholibegloo et al., 2018). The main limitations of these techniques are use of the appropriate photosensitizer and the light source.

This technique is being used in the control of biofilm related dental infections. A study has shown the removal of almost 95% of the biofilm on teeth as well as tooth whitening (Zhang et al., 2021). In another study, this approach along with the use of nanoparticles have been found effective for elimination of infectious bacteria and eradication of bacterial biofilms in the cornea. Bacterial biofilm causing infectious keratitis may leads to corneal blindness, however, matrix metalloproteinase (MMP)-sensitive supramolecular nanoparticles have been shown to enhance the photodynamic antibacterial effect (Han et al., 2020).

In another clinical study, treatment of *S. aureus* infections has been reported using PDT, however, it has also been suggested the need for appropriate trials for optimal results and safety (Pérez et al., 2021). The superiority of this approach increases in clinical settings when current therapies in the form of surgical operations, radiations or chemotherapeutic approaches are not successful, especially for cancer. Further progress on finding appropriate molecules is also being made. For example, photosensitizers need to be localized to the target tissue, and the use of nanocarriers combined with photosensitizers can further enhance accumulation of photosensitizers in the tumor site (Park et al., 2021).

Reported photosensitizer molecules are methylene blue, toluidine blue, 23H-porphine, tetrakis (1-methyl-pyridino)-21H, tetrakis (phenylthio)-29H, 31H-phthalocyanine, and tetra-p-tosylate salt (Sadekuzzaman et al., 2015; Srinivasan et al., 2021). Methylene blue-based photodynamic therapy has been reported for the treatment of antibiotic-resistant biofilms, antibiotic-resistant wound infections, and azole-resistant oral candidiasis (Hu et al., 2018). However, there are limitations to most of the first-generation photosensitizer molecules such as aggregation in water production, low levels of free radicle products, high dose demand and long half-life elimination. Further attempts have to be made to establish the better knowledge of the detailed mechanism of action for successful photodynamic therapy in eradicating bacterial biofilm formation (Park et al., 2021).

Cold Atmospheric Plasma (CAP)

Another promising therapeutic technique used to control bacterial biofilm infections is the use of cold atmospheric plasma (CAP). This involves a chemically diverse mixtures of reactive species and intermediates (ions and electrons, radical species, molecules in the ground or excited state and quanta of electromagnetic radiation (UV photons and visible light) and their interaction with a biofilm extracellular polymeric matrix (EPS) and cells (Gilmore et al., 2018).

CAP shows antibiofilm response by causing physical and molecular damage to the cell by DNA modification, protein inactivation and bactericidal effects (Lackmann et al., 2013). Death of *Candida albicans* occurs as a result of the destruction of the outer structure (Puligundla et al., 2017; Shi et al., 2009). Recently, CAP technology has been used in various clinical applications without damaging surrounding tissues (Isbary et al., 2013). Among many nosocomial infections, Ventilator-associated pneumonia (VAP) has high rates of mortality and morbidity rates caused by *P. aeruginosa, Acinetobacter baumannii, S. aureus*, and *C. albicans* biofilms Two different plasma treatments effectively inhibit such biofilms by damaging bacterial cell membranes without damaging human tissue (İbiş et al., 2020).

A Cold atmospheric pressure plasma (CAPP) has been used in a recent study as a novel approach to inactivate biofilms on hospital surfaces. Decontamination was most promising when detergent is used together with CAPP (Fallon et al., 2021).

Enzymatic Approach

Bacterial biofilms can also be controlled by controlling the production of extracellular matrix material that has variable composition depending on bacterial species. The dispersion of bacterial biofilm EPS by the use of different enzymes has been an effective strategy in the control of biofilms (Liaqat et al., 2021). These enzymes are grouped into three types depending on their target component of the biofilm matrix as: polysaccharide-degrading enzymes, nucleases and proteases (Li et al., 2017).

Different enzymes have the ability to degrade bacterial EPS. These enzymes have ability to inhibit the pathogenic ability of bacteria and can aid wound healing Three EPS degrading enzymes trypsin, β-glucosidase, and DNase I enzymes were found to be effective *at concentrations of* 1 μg/mL, 8 U/mL, and 150 U/mL, respectively in the control of biofilms developed by *P. aeruginosa* and *S. aureus, both commonly occurring species in wounds* (Fanaei et al., 2020). Similarly, the enzyme pyruvate dehydrogenase (PDH) has been reported to reduce 2.9 fold the biofilm formation of *P. aeruginosa* (Goodwine et al., 2019). Dispersin B and fructanases, from the class of extracellular glycosidases which mainly target the bacterial polysaccharides (Manuel et al., 2007) also have potential efficacy as therapeutic agents (Kaplan, 2009).

Dispersin B has the ability to remove already established biofilms from *A. actinomycetemcomitans* (Izano et al., 2007) and *S. epidermidis* (Izano et al., 2008; Kaplan et al., 2004). Recently, these enzymes have been immobilized on magnetic Fe_3O_4 and SiO_2 particles to enhance their efficacy to control biofilms in medical applications with less toxic effects (Liu et al., 2021). Dispersin B has been used in combination with different compounds to reduce biofilm formation of pathogenic bacterial species. For example, when used in combination with triclosan it will control the biofilm of *S. aureus*, *S. epidermidis*, and *E. coli* (Darouiche et al., 2009). Another group of enzymes known as "Deoxyribonuclease I" can digest eDNA with biofilm dispersing activity in pathogenic species of *Staphylococcus*, *P. aeruginosa*, *A. baumannii*, *E. coli*, *Haemophilus influenzae*, and *K. pneumoniae* (Kaplan et al., 2012; Tetz et al., 2009).

Cleaving biofilm matrix proteins by the enzyme proteinase K is another adjunctive therapeutic tool to inhibit oral biofilms (Liaqat et al., 2021; Karygianni et al., 2021). Dispersin B when used with protease (proteinase K or trypsin) has been reported to eliminate biofilms of many pathogenic staphylococcal strains on inert surfaces (Chaignon et al., 2007). The high-temperature-requiring protein A (HtrA) family of proteases are prerequisite for bacterial pathogenesis and survival in stress conditions. These proteins are required for lysing epithelial cell barriers and can be inhibited by protein inhibitors A statistically significant plaque biofilm removal (10%) was recorded when lysozyme (2500 μg/mL) were tested on Peri-implantitis as biofilm destabilizing agents (Matthes et al., 2021). Other enzymes reported to have biofilm removal efficacy are α-amylase having efficacy to remove *S. aureus* biofilms (Craigen et al., 2011), lyase (20 μg/mL) and gentamycin (64 μg/mL) against two mucoid *P. aeruginosa species* and Lactonase against four *P. aeruginosa* strains (Alkawash et al., 2006; Sadekuzzaman et al., 2015). While the use of enzymes for biofilm removal and eradication is an emerging approach, and mixtures of different enzymes can improve biofilm eradication efficiency with reference to different bacterial species, these factors need to be optimized and are challenging.

Surfactants

There have been a number of different medically acceptable chemicals with the ability to clean wounds and combat biofilms. There is a search for new surfactants with high biofilm inhibition besides excellent cellular repair. Compounds such as betaines and poloxamers have the ability to moisturize and smooth to the skin besides removing cellular debris and preventing biofilm formation (Belfield et al., 2017; Yang et al., 2017).

These compounds have hydrophobic and hydrophilic groups and have broad spectrum antibiotic activity besides debris removal efficiency (Percival et al., 2019). The significance of these biosurfactants have been enhanced during the COVID-19 pandemic where effective biosurfactant ingredients have been investigated for effective cleaning formulas and surface coatings as well as for drug delivery system for combatting SARS-CoV-2 symptoms (Ceresa et al., 2021).

Recently, betaine surfactant and polyhexanide have been found to control biofilm in cochlear implants. Results show that their use significantly reduces morbidity in surgery (Suri et al., 2021). In one study, efficacy of biosurfactants known as rhamnolipids are effective enough to control biofilms by inhibiting bacterial cells, virulence (Chrzanowski et al., 2012) biofilm formation (Fracchia et al., 2012) and affecting bacterial cell wall charge (Walencka et al., 2008). In a study where 1% silver sulfadiazinethe surfactant-based wound dressing was tested, efficient biofilm detachment *in vitro* was found for *P. aeruginosa*, *Enterococcus* sp., *S. epidermidis* and *S. aureus* (Percival et al., 2017). Surfactants have been reported to be used alone or as an adjunct possible therapy for the treatment of hard-to-treat chronic rhinosinusitis. However, surfactants may lead to nasal irritation which may limit their use (Rosen et al., 2013).

Surfactants have also been reported to modify the surface properties and inhibit bacterial adherence to different surfaces (Rodrigues et al., 2004; Sadekuzzaman et al., 2015; Walencka et al., 2008). Potential surfactants alone or as adjunct therapy may have desloughing properties that helps to clean and heal wounds; however, *in vivo* efficacy is still a challenge that requires to be proven in clinical trials.

Surface Modification

Another strategy for the control or prevention of biofilms on surfaces is to change the surface properties. For example incorporation of organic and inorganic (Nichols, 2005; Vitiello et al., 2018) and cationic biocides on to surfaces (Carmona-Ribeiro et al., 2013) can be potent enough to discourage biofilm formation.

The antibiofilm properties of PA-11 modified with the polymeric biocide PHMG-NS show antibiofilm efficacy against opportunistic biofilm-forming *E. coli* K12 and *S. aureus* ATCC 25923 and thus have potential to reduce the antibiotic resistance in clinical setups. Another biocide with antibiofilm activity is Goldshield (GS5) that has been reported to be efficient against *S. aureus* ATCC43300 with longer-term bactericidal activity against *S. aureus*, and was found to be efficient with reference to infection control measures (Murray et al., 2017).

In vitro anti-biofilm activity has been shown for benzalkonium chloride (BZC), chlorhexidine digluconate (CHX), and triclosan (TRC) against three species of Candida (*C. albicans*, *C. parapsilosis*, and *C. glabrata*) (Ozturk et al., 2020). LH IDROXI FAST and LH ENZYCLEAN SPRAY biocides show promising antibiofilm action on biofilm forming *E. hirae* biomasses (Di Lodovico et al., 2017).

Another strategy to prevent biofilm formation in medical settings is the use of nanomaterials as surface coatings on clinical implantable tools and devices in hospitals. Their biocompatibility and novel composition make them suitable for medical devices. Some reported examples of use are on contact lenses, oral implants, heart valves, catheters in UTI, artificial joints, etc. (Ramasamy et al., 2016). With their small sizes and large surface to volume ratio, their penetration power into biofilms and bacterial cells results in cell damage and makes them valuable entities as drugs delivery system besides targeting entities.

Nanosized TiO_2 are effective enough showing long term antibiofilm and antifungal efficacy especially on medical devices when compared to other metallic oxides. Their application as self-cleaning and antifogging agents have already been utilized (Ramasamy and Lee, 2016).

Catheters used in hospitals when coated with iron nanoparticles have promising antibiofilm efficacy against *S. aureus* and *P. aeruginosa*, and, being nontoxic, there *in vivo* applications are recommended. (Anghel et al., 2012). Silver nanoparticles (Ag NPs) have been found promising in biofilm inhibition of clinically significant pathogens *P. aeruginosa*, *E. coli*, and *S. aureus* without showing cytotoxicity to mammalian cells (Yi et al., 2022).

Nanofabrication of implantable biomaterials has a great potential application to control infections on medical devices. Barium titanate (BTO)-based composite materials have not only been used in biological applications but also reported with biocompatibility toward human gingival fibroblasts and keratinocytes. Overall, BTO-composites are promising as antibiofilm surfaces (Dhall et al., 2021).

Vetiveria zizanioides root extract-mediated silver nanoparticles (AgNPs) are reported to show potent antibiofilm and anti-quorum sensing against challenging medical and nosocomial infections caused by *S. marcescens* (Ravindran et al., 2018). Devices engineered with polymeric NPs ensure antibacterial potential by releasing antibiotics and bacteriostatic proteins that ensure killing of bacterial cells and biofilms. Although nanoparticles and biocides have remarkable properties, these same properties may cause some limitations that need to be addressed like more surface area with non-specificity of the nanoparticles antimicrobial efficacy, non-targeted biocides and random DNA mutations on both pathogenic and symbiotic microbial flora. Similarly, the toxicity induced by chemically synthesized nanoparticles and biocides is also of concern (Cappitelli et al., 2021; Shkodenko et al., 2020)

Bacterial pathogenesis in the form of biofilms increases when a bacterial cell comes into contact with any surface. With the involvement of van der Waals and electrostatic forces, bacterial cell attachment is triggered (Krsmanovic et al., 2021). Various coatings have been developed with the properties to modulate biofilm development for example hydrogel systems (poly ethylene glycol with sequestered antibiotic ciprofloxacin) that reduces cell adhesion to medical tools (silicone catheters) in *in vitro* assays against *P. aeruginosa (Park et al., 1998)*.

Development of more advanced bioactive and biocidal thermoplastic PU materials is on the horizon. These will have high antibacterial activity to address the existing challenges in biomedical applications such as biomedical devices (catheters) and wound dressing (Gharibi et al., 2021). Association of Chitosan with different enzymes and antimicrobial peptides enable them to be used as antimicrobial materials for the development of new coating surfaces in field of clinical science (Teixeira-Santos et al., 2021). Superhydrophobic surfaces have been developed offering antifouling and antibiofilm properties for medical implants where they can hinder biofilm formation (Shahid et al., 2019).

Conclusion

In this chapter, we have reviewed the detailed development of biofilm control in clinical and medical settings with their socioeconomical significance. Significant research has been done on novel systems besides using traditional approaches of antibiotic therapy for biofilm removal and eradication. The chapter also highlighted future horizons for the improvement of current technologies. Approaches have been discussed through which antibiotic resistance was not developed in bacterial cells so they seem promising for the control of biofilm-based infections in healthcare settings. Many advancements and new developments in this area are yet to be explored with broad level of *in vivo* and *in vitro* clinical trials. However, development of advanced technologies with less cost and pain without compromising antibiotic resistance development (quorum quenching and other techniques) will be helpful to address the social issue of inequity of access to healthcare facilities especially in case of the currently prevailing global crisis of COVID-19.

References

A Sharqi, H. and M Al-Rawi, A. (2017). Detection of Bacteria Causing Conjunctivitis in Patients Using Lenses for Vision Correction. *Rafidain Journal of Science*, 44-52.

Afonso, T. B., Simões, L. C. and Lima, N. (2021). Methylobacterium oryzae influences isoepoxydon dehydrogenase gene expression and patulin production by *Penicillium expansum. Water*, 1427.

Alexpandi, R., Abirami, G., Satish, L., Swasthikka, R. P., Krishnaveni, N., Jayakumar, R., Pandian, S. K. and Ravi, A. V. (2021). Tocopherol and phytol possess anti-quorum sensing mediated anti-infective behavior against *Vibrio campbellii* in aquaculture: An *in vitro* and *in vivo* study. *Microbial Pathogenesis*, 105221.

Alkawash, M. A., Soothill, J. S. and Schiller, N. L. (2006). Alginate lyase enhances antibiotic killing of mucoid Pseudomonas aeruginosa in biofilms. *Apmis*, 131-138.

Anghel, I., Limban, C., Grumezescu, A. M., Anghel, A. G., Bleotu, C. and Chifiriuc, M. C. (2012). *In vitro* evaluation of anti-pathogenic surface coating nanofluid, obtained by combining Fe 3 O 4/C 12 nanostructures and 2-((4-ethylphenoxy) methyl)-N-(substituted-phenylcarbamothioyl)-benzamides. *Nanoscale research letters*, 1-10.

Belfield, K., Bayston, R., Hajduk, N., Levell, G., Birchall, J. P. and Daniel, M. (2017). Evaluation of combinations of putative anti-biofilm agents and antibiotics to eradicate biofilms of *Staphylococcus aureus* and *Pseudomonas aeruginosa. Journal of Antimicrobial Chemotherapy*, 2531-2538.

Bertsch, P., Etter, D. and Fischer, P. (2021). Transient *in situ* measurement of kombucha biofilm growth and mechanical properties. *Food & function*, 4015-4020.

Bhatt, V.S. 2018. "Quorum sensing mechanisms in gram positive bacteria". In *Implication of Quorum Sensing System in Biofilm Formation and Virulence*., 297-311. Springer.

Cappitelli, F. and Villa, F. (2021). Novel Antibiofilm Non-Biocidal Strategies. *Microorganisms in the Deterioration and Preservation of Cultural Heritage*, 117.

Carmona-Ribeiro, A. M. and de Melo Carrasco, L. D. (2013). Cationic antimicrobial polymers and their assemblies. *International journal of molecular sciences*, 9906-9946.

Ceresa, C., Fracchia, L., Fedeli, E., Porta, C. and Banat, I. M. (2021). Recent advances in biomedical, therapeutic and pharmaceutical applications of microbial surfactants. *Pharmaceutics*, 466.

Chaignon, P., Sadovskaya, I., Ragunah, C., Ramasubbu, N., Kaplan, J. and Jabbouri, S. (2007). Susceptibility of staphylococcal biofilms to enzymatic treatments depends on their chemical composition. *Applied microbiology and biotechnology*, 125-132.

Chen, C., Krishnan, V., Macon, K., Manne, K., Narayana, S.V. and Schneewind, O. (2013a). Secreted proteases control autolysin-mediated biofilm growth of Staphylococcus aureus. *Journal of Biological Chemistry*, 29440-29452.

Chen, M., Yu, Q. and Sun, H. (2013b). Novel strategies for the prevention and treatment of biofilm related infections. *International journal of molecular sciences*, 18488-18501.

Choi, H., Mascuch, S. J., Villa, F. A., Byrum, T., Teasdale, M. E., Smith, J. E., Preskitt, L. B., Rowley, D. C., Gerwick, L. and Gerwick, W. H. (2012). Honaucins A– C, potent inhibitors of inflammation and bacterial quorum sensing: synthetic derivatives and structure-activity relationships. *Chemistry & biology*, 589-598.

Christensen, Q. H., Grove, T. L., Booker, S. J. and Greenberg, E. P. (2013). A high-throughput screen for quorum-sensing inhibitors that target acyl-homoserine lactone synthases. *Proceedings of the National Academy of Sciences*, 13815-13820.

Chrzanowski, Ł., Ławniczak, Ł. and Czaczyk, K. (2012). Why do microorganisms produce rhamnolipids? *World Journal of Microbiology and Biotechnology*, 401-419.

Craigen, B., Dashiff, A. and Kadouri, D. E. (2011). The use of commercially available alpha-amylase compounds to inhibit and remove *Staphylococcus aureus* biofilms. *The open microbiology journal*, 21.

Darmani, H., Tawalbeh, K. H., Al-Hiyasat, A. S. and Al-Akhras, M.-A. (2018). Comparison of the photosensitivity of biofilms of different genera of cariogenic bacteria in tooth slices. *Polish journal of microbiology*, 455.

Darouiche, R. O., Mansouri, M. D., Gawande, P. V. and Madhyastha, S. (2009). Antimicrobial and antibiofilm efficacy of triclosan and DispersinB® combination. *Journal of antimicrobial chemotherapy*, 88-93.

De Silva, E. H., Salamat, N., Zhang, L., Zheng, J. and Novak, B. M. (2021). Water-soluble polycarbodiimides and their cytotoxic and antifungal properties. *Journal of Biomaterials Science*, 1-18.

Dewasthale, S., Mani, I. and Vasdev, K. (2018). Microbial biofilm: current challenges in health care industry. *Journal of Applied Biotechnology & Bioengineering*, 5,

Dhall, A., Islam, S., Park, M., Zhang, Y., Kim, A. and Hwang, G. (2021). Bimodal Nanocomposite Platform with Antibiofilm and Self-Powering Functionalities for Biomedical Applications. *ACS Applied Materials & Interfaces*, 40379-40391.

Di Lodovico, S., Cataldi, V., Di Campli, E., Ancarani, E., Cellini, L. and Di Giulio, M. (2017). *Enterococcus hirae* biofilm formation on hospital material surfaces and effect of new biocides. *Environmental health and preventive medicine*, 1-7.

Dong, Y.-H., Xu, J.-L., Li, X.-Z. and Zhang, L.-H. (2000). AiiA, an enzyme that inactivates the acylhomoserine lactone quorum-sensing signal and attenuates the virulence of *Erwinia carotovora*. *Proceedings of the National Academy of Sciences*, 3526-3531.

Eick, S. (2021). Biofilms. *Oral Biofilms*, 1-11.

Evans, L. V. 2000. *Biofilms: recent advances in their study and control.* CRC press.

Fallon, M., Kennedy, S., Kumar, S., Daniels, S. and Humphreys, H. (2021). The Potential Use of a Cold Atmospheric Plasma Jet for Decontamination of Hospital Surfaces. A Pilot Study. *Plasma Medicine*, 11,

Fanaei Pirlar, R., Emaneini, M., Beigverdi, R., Banar, M., B. van Leeuwen, W. and Jabalameli, F. (2020). Combinatorial effects of antibiotics and enzymes against dual-species *Staphylococcus aureus* and *Pseudomonas aeruginosa* biofilms in the wound-like medium. *Plos one*, e0235093.

Fernández-Hidalgo, N. and Almirante, B. (2014). Antibiotic-lock therapy: a clinical viewpoint. *Expert review of anti-infective therapy*, 117-129.

Fracchia, L., Cavallo, M., Martinotti, M. G. and Banat, I. M. (2012). Biosurfactants and bioemulsifiers biomedical and related applications–present status and future potentials. *Biomedical science, engineering and technology*, 326-335.

Gharibi, R. and Agarwal, S. (2021). Polyurethanes from Hydrophobic Elastic Materials to Hydrogels with Potent Nonleaching Biocidal and Antibiofilm Activity. *ACS Applied Polymer Materials*, 4695-4707.

Ghavamian, S., Hay, I. D., Habibi, R., Lithgow, T. and Cadarso, V. J. (2021). Three-Dimensional Micropatterning Deters Early Bacterial Adherence and Can Eliminate Colonization. *ACS Applied Materials & Interfaces*.

Gholibegloo, E., Karbasi, A., Pourhajibagher, M., Chiniforush, N., Ramazani, A., Akbari, T., Bahador, A. and Khoobi, M. (2018). Carnosine-graphene oxide conjugates decorated with hydroxyapatite as promising nanocarrier for ICG loading with enhanced antibacterial effects in photodynamic therapy against *Streptococcus mutans*. *Journal of Photochemistry and Photobiology B: Biology*, 14-22.

Gilmore, B. F., Flynn, P. B., O'Brien, S., Hickok, N., Freeman, T. and Bourke, P. (2018). Cold plasmas for biofilm control: opportunities and challenges. *Trends in biotechnology*, 627-638.

Goodwine, J., Gil, J., Doiron, A., Valdes, J., Solis, M., Higa, A., Davis, S. and Sauer, K. (2019). Pyruvate-depleting conditions induce biofilm dispersion and enhance the efficacy of antibiotics in killing biofilms *in vitro* and *in vivo*. *Scientific reports*, 1-16.

Halaji, M., Karimi, A., Shoaei, P., Nahaei, M., Khorvash, F., Ataei, B., Yaran, M. and Havaei, S. A. (2017). Distribution of SCCmec elements and presence of Panton-Valentine Leukocidin in Methicillin-Resistant *Staphylococcus epidermidis* isolated from clinical samples in a University Hospital of Isfahan City, Iran. *Journal of clinical and diagnostic research: JCDR*, DC27.

Han, H., Gao, Y., Chai, M., Zhang, X., Liu, S., Huang, Y., Jin, Q., Grzybowski, A., Ji, J. and Yao, K. (2020). Biofilm microenvironment activated supramolecular nanoparticles for enhanced photodynamic therapy of bacterial keratitis. *Journal of Controlled Release*, 676-687.

Hengzhuang, W., Wu, H., Ciofu, O., Song, Z. and Høiby, N. (2011). Pharmacokinetics/pharmacodynamics of colistin and imipenem on mucoid and nonmucoid Pseudomonas aeruginosa biofilms. *Antimicrobial agents and chemotherapy*, 4469-4474.

Høiby, N., Ciofu, O., Johansen, H. K., Song, Z. j., Moser, C., Jensen, P. Ø., Molin, S., Givskov, M., Tolker Nielsen, T. and Bjarnsholt, T. (2011). The clinical impact of bacterial biofilms. *International journal of oral science*, 55-65.

Hu, X., Huang, Y.-Y., Wang, Y., Wang, X. and Hamblin, M.R. (2018). Antimicrobial photodynamic therapy to control clinically relevant biofilm infections. *Frontiers in microbiology*, 1299.

Hume, E., Baveja, J., Muir, B., Schubert, T., Kumar, N., Kjelleberg, S., Griesser, H., Thissen, H., Read, R. and Poole-Warren, L. (2004). The control of *Staphylococcus epidermidis* biofilm formation and *in vivo* infection rates by covalently bound furanones. *Biomaterials*, 5023-5030.

Hussain, M., Wilcox, M. and White, P. (1993). The slime of coagulase-negative staphylococci: biochemistry and relation to adherence. *FEMS microbiology reviews*, 191-208.

İbiş, F. and Ercan, U. K. (2020). Inactivation of biofilms in endotracheal tube by cold atmospheric plasma treatment for control and prevention of ventilator associated pneumonia. *Plasma Processes and Polymers*, 2000065.

Isbary, G., Shimizu, T., Li, Y.-F., Stolz, W., Thomas, H. M., Morfill, G. E. and Zimmermann, J. L. (2013). Cold atmospheric plasma devices for medical issues. *Expert review of medical devices*, 367-377.

Izano, E., Wang, H., Ragunath, C., Ramasubbu, N. and Kaplan, J. (2007). Detachment and killing of *Aggregatibacter actinomycetemcomitans* biofilms by dispersin B and SDS. *Journal of dental research*, 618-622.

Izano, E. A., Amarante, M. A., Kher, W. B. and Kaplan, J. B. (2008). Differential roles of poly-N-acetylglucosamine surface polysaccharide and extracellular DNA in *Staphylococcus aureus* and *Staphylococcus epidermidis* biofilms. *Applied and environmental microbiology*, 470-476.

Jain, R. and Danziger, L. H. (2004). Multidrug-resistant Acinetobacter infections: an emerging challenge to clinicians. *Annals of Pharmacotherapy*, 1449-1459.

Javanshir, N., Hosseini, G. N. G., Sadeghi, M., Esmaeili, R., Satarikia, F., Ahmadian, G. and Allahyari, N. (2021). Evaluation of the Function of Probiotics, Emphasizing the Role of their Binding to the Intestinal Epithelium in the Stability and their Effects on the Immune System. *Biological Procedures Online*, 1-17.

Josse, J., Valour, F., Maali, Y., Diot, A., Batailler, C., Ferry, T. and Laurent, F. (2019). Interaction between staphylococcal biofilm and bone: How does the presence of biofilm promote prosthesis loosening? *Frontiers in microbiology*, 1602.

Kaplan, J.B. (2009). Therapeutic potential of biofilm-dispersing enzymes. *The International journal of artificial organs*, 545-554.

Kaplan, J.B., LoVetri, K., Cardona, S.T., Madhyastha, S., Sadovskaya, I., Jabbouri, S. and Izano, E.A. (2012). Recombinant human DNase I decreases biofilm and increases antimicrobial susceptibility in staphylococci. *The Journal of antibiotics*, 73-77.

Kaplan, J.B., Ragunath, C., Velliyagounder, K., Fine, D.H. and Ramasubbu, N. (2004). Enzymatic detachment of *Staphylococcus epidermidis* biofilms. *Antimicrobial agents and chemotherapy*, 2633-2636.

Karygianni, L., Paqué, P.N., Attin, T. and Thurnheer, T. (2021). Single DNase or Proteinase Treatment Induces Change in Composition and Structural Integrity of Multispecies Oral Biofilms. *Antibiotics*, 400.

Krsmanovic, M., Biswas, D., Ali, H., Kumar, A., Ghosh, R. and Dickerson, A. K. (2021). Hydrodynamics and surface properties influence biofilm proliferation. *Advances in Colloid and Interface Science*, 102336.

Lackmann, J.-W., Schneider, S., Edengeiser, E., Jarzina, F., Brinckmann, S., Steinborn, E., Havenith, M., Benedikt, J. and Bandow, J. E. (2013). Photons and particles emitted from cold atmospheric-pressure plasma inactivate bacteria and biomolecules independently and synergistically. *Journal of the Royal Society Interface*, 20130591.

Lahiri, D., Nag, M., Sheikh, H. I., Sarkar, T., Edinur, H. A., Pati, S. and Ray, R. R. (2021). Microbiologically-synthesized nanoparticles and their role in silencing the biofilm signaling cascade. *Frontiers in microbiology*, 180.

Lee, S.W., Phillips, K.S., Gu, H., Kazemzadeh-Narbat, M. and Ren, D. (2021). How microbes read the map: Effects of implant topography on bacterial adhesion and biofilm formation. *Biomaterials*, 120595.

Li, X.-H. and Lee, J.-H. (2017). Antibiofilm agents: A new perspective for antimicrobial strategy. *Journal of Microbiology*, 753-766.

Li, Y.-H., Lau, P.C., Lee, J. H., Ellen, R. P. and Cvitkovitch, D. G. (2001). Natural genetic transformation of *Streptococcus mutans* growing in biofilms. *Journal of bacteriology*, 897-908.

Liaqat, I., Bachmann, R.T., Nasim Sabri, A., Edyvean, R.G. and Biggs, C.A. (2008). Investigating the effect of patulin, penicillic acid and EDTA on biofilm formation of isolates from dental unit water lines. *Applied microbiology and biotechnology*, 349-358.

Liaqat, I. and Sabri, A.N. (2008). Effect of biocides on biofilm bacteria from dental unit water lines. *Current microbiology*, 619-624.

Liaqat, I., Hussain, T., Qurashi, A.W., Saleem, G., Bibi, A., Qamar, M.F. and Ali, S. (2021). Antibiofilm Activity of Proteolytic Enzymes against Salmonella Gallinarum Isolates from Commercial Broiler Chickens. *Pakistan Journal of Zoology*, 1111.

Liu, Z., Zhao, Z., Zeng, K., Xia, Y., Xu, W., Wang, R., Guo, J. and Xie, H. (2021). Functional Immobilization of a Biofilm-Releasing Glycoside Hydrolase Dispersin B on Magnetic Nanoparticles. *Applied Biochemistry and Biotechnology*, 1-11.

Manuel, S. G., Ragunath, C., Sait, H. B., Izano, E. A., Kaplan, J. B. and Ramasubbu, N. (2007). Role of active site residues of dispersin B, a biofilm releasing β hexosaminidase from a periodontal pathogen, in substrate hydrolysis. *The FEBS journal*, 5987-5999.

Mare, A. D., Man, A., Ciurea, C. N., Toma, F., Cighir, A., Mareș, M., Berța, L. and Tanase, C. (2021). Silver Nanoparticles Biosynthesized with Spruce Bark Extract—A Molecular Aggregate with Antifungal Activity against Candida Species. *Antibiotics*, 1261.

Matthes, R., Jablonowski, L., Holtfreter, B., Pink, C. and Kocher, T. (2021). Enzymatic biofilm destabilisation to support mechanical cleansing of inserted dental implant surfaces: an in-vitro pilot study. *Odontology*, 1-12.

Murray, J., Muruko, T., Gill, C. I., Kearney, M. P., Farren, D., Scott, M. G., McMullan, G. and Ternan, N. G. (2017). Evaluation of bactericidal and anti-biofilm properties of a novel surface-active organosilane biocide against healthcare associated pathogens and *Pseudomonas aeruginosa* biolfilm. *PLoS One*, e0182624.

Nichols, D. 2005. *Biocides in plastics.* iSmithers Rapra Publishing

Nielsen, P. H., Jahn, A. and Palmgren, R. (1997). Conceptual model for production and composition of exopolymers in biofilms. *Water Science and Technology*, 11-19.

Nwoko, E.-s.Q. and Okeke, I. N. (2021). Bacteria autoaggregation: how and why bacteria stick together. *Biochemical Society Transactions*.

Ozturk, A., Abdulmajed, O. and Aydin, M. (2020). Investigation of antifungal, antibiofilm and anti-filamentation activities of biocides against *Candida* isolates. *Annals of Medical Research*, 2041-2046.

Pachori, P., Gothalwal, R. and Gandhi, P. (2019). Emergence of antibiotic resistance *Pseudomonas aeruginosa* in intensive care unit; a critical review. *Genes & diseases*, 109-119.

Park, J., Lee, Y.-K., Park, I.-K. and Hwang, S. R. (2021). Current Limitations and Recent Progress in Nanomedicine for Clinically Available Photodynamic Therapy. *Biomedicines*, 85.

Park, K. D., Kim, Y. S., Han, D. K., Kim, Y. H., Lee, E. H. B., Suh, H. and Choi, K. S. (1998). Bacterial adhesion on PEG modified polyurethane surfaces. *Biomaterials*, 851-859.

Paul, D., Kim, Y. S., Ponnusamy, K. and Kweon, J. H. (2009). Application of quorum quenching to inhibit biofilm formation. *Environmental Engineering Science*, 1319-1324.

Percival, S. L., Mayer, D., Kirsner, R. S., Schultz, G., Weir, D., Roy, S., Alavi, A. and Romanelli, M. (2019). Surfactants: role in biofilm management and cellular behaviour. *International wound journal*, 753-760.

Percival, S.L., Mayer, D. and Salisbury, A.M. (2017). Efficacy of a surfactant-based wound dressing on biofilm control. *Wound Repair and Regeneration*, 767-773.

Pérez, C., Zúñiga, T. and Palavecino, C. E. (2021). Photodynamic Therapy for Treating *Staphylococcus aureus* Infections. *Photodiagnosis and Photodynamic Therapy*, 102285.

Puligundla, P. and Mok, C. (2017). Potential applications of nonthermal plasmas against biofilm associated micro organisms *in vitro*. *Journal of applied microbiology*, 1134-1148.

Qurashi, A. W. and Sabri, A. N. (2012a). Bacterial exopolysaccharide and biofilm formation stimulate chickpea growth and soil aggregation under salt stress. *Brazilian Journal of Microbiology*, 1183-1191.

Qurashi, A. W. and Sabri, A. N. (2012b). Biofilm formation in moderately halophilic bacteria is influenced by varying salinity levels. *Journal of basic microbiology*, 566-572.

Rafi, U., Asmat, M. N., Bokhari, S. S. and Qurashi, A. W. (2019). Biofilm Formation and Bacterial Aggregation Response of *Planomicrobium chinense* and *Alkaligenes faecalis* Associated with Periplaneta americana Found in Household Sewerage. *Pakistan Journal of Zoology*, 2389-2392.

Ramasamy, M. and Lee, J. (2016). Recent nanotechnology approaches for prevention and treatment of biofilm-associated infections on medical devices. *BioMed Research International*, 2016,

Rasmussen, T. B. and Givskov, M. (2006). Quorum sensing inhibitors: a bargain of effects. *Microbiology*, 895-904.

Ravindran, D., Ramanathan, S., Arunachalam, K., Jeyaraj, G., Shunmugiah, K. and Arumugam, V. (2018). Phytosynthesized silver nanoparticles as antiquorum sensing and antibiofilm agent against the nosocomial pathogen *Serratia marcescens*: an *in vitro* study. *Journal of applied microbiology*, 1425-1440.

Rodrigues, L., Van der Mei, H., Teixeira, J. A. and Oliveira, R. (2004). Biosurfactant from *Lactococcus lactis* 53 inhibits microbial adhesion on silicone rubber. *Applied microbiology and biotechnology*, 306-311.

Rogers, K. L., Fey, P. D. and Rupp, M. E. (2009). Coagulase-negative staphylococcal infections. *Infectious disease clinics of North America*, 73-98.

Rosen, P. L., Palmer, J. N., O'Malley Jr, B. W. and Cohen, N. A. (2013). Surfactants in the management of rhinopathologies. *American journal of rhinology & allergy*, 177-180.

Sadekuzzaman, M., Yang, S., Mizan, M. and Ha, S. (2015). Current and recent advanced strategies for combating biofilms. *Comprehensive Reviews in Food Science and Food Safety*, 491-509.

Schilcher, K. and Horswill, A. R. (2020). Staphylococcal biofilm development: structure, regulation, and treatment strategies. *Microbiology and Molecular Biology Reviews*, 84, e00026-00019.

Shahid, A., Rasool, M., Akhter, N., Aslam, B., Hassan, A., Sana, S., Rasool, M. H. and Khurshid, M. 2019. "Innovative strategies for the control of biofilm formation in clinical settings". In *Bacterial Biofilms*. IntechOpen London, UK.

Shi, X. and Zhu, X. (2009). Biofilm formation and food safety in food industries. *Trends in Food Science & Technology*, 407-413.

Shkodenko, L., Kassirov, I. and Koshel, E. (2020). Metal oxide nanoparticles against bacterial biofilms: perspectives and limitations. *Microorganisms*, 1545.

Song, Y., Ke, Y., Kang, M. and Bao, R. (2021). Function, molecular mechanisms, and therapeutic potential of bacterial HtrA proteins: An evolving view. *Computational and Structural Biotechnology Journal*.

Srinivasan, R., Santhakumari, S., Poonguzhali, P., Geetha, M., Dyavaiah, M. and Xiangmin, L. (2021). Bacterial biofilm inhibition: A focused review on recent therapeutic strategies for combating the biofilm mediated infections. *Frontiers in Microbiology*, 1106.

Subhadra, B., Kim, D. H., Woo, K., Surendran, S. and Choi, C. H. (2018). Control of biofilm formation in healthcare: Recent advances exploiting quorum-sensing interference strategies and multidrug efflux pump inhibitors. *Materials*, 1676.

Suri, N., Yadav, C., Sandilya, S. and Bhalodia, N. (2021). Salvaging Cochlear Implant After Suspected Biofilm Infection: Our Experience. *Indian Journal of Otolaryngology and Head & Neck Surgery*, 1-5.

Sutherland, I. W. (2001). Biofilm exopolysaccharides: a strong and sticky framework. *Microbiology*, 3-9.

Tan, M., Lau, J. and Guglielmo, B. J. (2014). Ethanol locks in the prevention and treatment of catheter-related bloodstream infections. *Annals of Pharmacotherapy*, 607-615.

Teixeira-Santos, R., Lima, M., Gomes, L. C. and Mergulhão, F. J. (2021). Antimicrobial coatings based on chitosan to prevent implant-associated infections: *A systematic review. iScience*, 103480.

Tetz, G. V., Artemenko, N. K. and Tetz, V. V. (2009). Effect of DNase and antibiotics on biofilm characteristics. *Antimicrobial agents and chemotherapy*, 1204-1209.

Vitiello, G., Silvestri, B. and Luciani, G. (2018). Learning from nature: bioinspired strategies towards antimicrobial nanostructured systems. *Current topics in medicinal chemistry*, 22-41.

Walencka, E., Różalska, S., Sadowska, B. and Różalska, B. (2008). The influence of Lactobacillus acidophilus-derived surfactants on staphylococcal adhesion and biofilm formation. *Folia microbiologica*, 61.

Wu, H., Moser, C., Wang, H.-Z., Høiby, N. and Song, Z.-J. (2015). Strategies for combating bacterial biofilm infections. *International journal of oral science*, 1-7.

Xiao, R. and Zheng, Y. (2016). Overview of microalgal extracellular polymeric substances (EPS) and their applications. *Biotechnology Advances*, 1225-1244.

Xue, R.-Y., Liu, C., Xiao, Q.-T., Sun, S., Zou, Q.-M. and Li, H.-B. (2021). HtrA-family proteases of bacterial pathogens: pros and cons for their therapeutic use. *Clinical Microbiology and Infection.*

Yang, Q., Larose, C., Della Porta, A. C., Schultz, G. S. and Gibson, D. J. (2017). A surfactant-based wound dressing can reduce bacterial biofilms in a porcine skin explant model. *International wound journal*, 408-413.

Yi, S.-Y., Lin, H.-Y., Yang, L.-C., Tseng, S.-C., Sun, A. Y., Chen, C. and Wan, D. (2022). Substrate-independent adsorption of nanoparticles as anti-biofilm coatings. *Biomaterials Science*.

Zare, M., Zare, M., Butler, J. A. and Ramakrishna, S. (2021). Nanoscience-Led Antimicrobial Surface Engineering to Prevent Infections. *ACS Applied Nano Materials*.

Zhang, H., Zhu, Y., Li, Y., Qi, X., Yang, J., Qi, H., Li, Q., Ma, Y., Zhang, Y. and Zhang, X. (2021). A Bifunctional Zwitterion Modified Porphyrin for Photodynamic Nondestructive Tooth Whitening and Biofilm Eradication. *Advanced Functional Materials*, 2104799.

Chapter 4

Major Physical Strategies for Controlling Biofilm Formation

Beenish Sarfraz[1], Aisha Waheed Qurashi[1], Sumaira Mazhar[1], Aisha Sultan[2] and Iram Liaqat[3,*]

[1]Department of Biology, Lahore Garrison University, Lahore, Pakistan
[2]Punjab Rangers Teaching Hospital, Lahore, Pakistan
[3]Microbiology Lab, Department of Zoology, Government College University, Lahore, Pakistan

Abstract

Biofilms are complex microbial communities found in a multitude of habitats. These sedentary accumulation of microbes reside on various kinds of surfaces from simple household pipework to medical devices like urinary catheters and even on plant and animal tissues. An extracellular matrix secreted by these biofilms make them more resistant towards conventional antibacterial therapies. Biofilms are responsible for causing various infections which are becoming difficult to treat due to ever increasing antibiotic resistance in bacterial species compounded by the resistance of the biofilm structure. Many antibiofilm strategies have been explored to tackle their incidence especially physical methods. This chapter will explore the use of methods to physically remove of biofilms from hospital related devices. Such methods include irradiation, ultrasound power, electric and magnetic fields. The mechanism of anti-biofilm action is discussed along with the effectiveness of combinations of treatment. Physical methods have demonstrated remarkable results which could help to minimize the incidence of biofilms.

---

*Corresponding Author's Email: iramliaq@hotmail.com.

In: Understanding Antibiofilm Activity
Editor: Iram Liaqat
ISBN: 978-1-68507-927-7

Keywords: biofilm, ultrasound, bacteria, electric filed, ultraviolet radiations

Introduction

Biofilms are formed by the amalgamation of microbes onto a surface, which results in a film or a layer formation (Costerton, 1999, Flemming et al., 2010). Microbes belonging to bacteria, fungi, and protists are naturally equipped with bio-machinery to make these unique and resistant films on any natural or artificial surfaces including rocks, ponds, ships, food, medical instruments, animal and plant tissues (Costerton et al., 1995). History has witnessed many outbreaks of pathogens due to biofilms and approximately 80% of microbial infections now a days are due to biofilms (Souza et al., 2015, Trancassini et al., 2014). Bacteria within these biofilms are more resistant (10–1000 times) to the action of typical antibiotics, disinfectants, and sanitizers than planktonic bacteria. Multiple theories have been proposed to explain greater resistance of biofilms. One of the reasons is the production of extracellular polymeric substances (EPS) which leads to less diffusion rate of antimicrobial substances (Limoli et al., 2015). Additionally, enhanced enzyme protection, less rate of bacterial respiration, and physiological changes in the organisms are also some of the reasons behind biofilm's increased persistence on different surfaces (Sharahi et al., 2019, Yuan et al., 2020).

Biofilm formation is a dynamic process and different mechanisms are involved in their attachment and growth. EPS plays an important role in the attachment and colonization of microorganisms to different surfaces (Flemming et al., 2016, Watnick et al., 2000). Biofilm environment is so dynamic, that they can physically and chemically change their microenvironment. They release different signaling molecules that help in intercellular communication. The quorum sensing (QS) process can alter gene expression by changes in cellular density. Owing to the complex environment of biofilms, they are quite resistant to external pressures like antimicrobial compounds, temperature, and pH conditions (Bridier et al., 2015). Treatment with biocides sometimes proves ineffective and they continue to thrive on different surfaces. Multiple chemical treatments from strong synthetic antimicrobial compounds to natural agents have been employed to control biofilms. Some of them proved effective but their resistance is ever-evolving against common biocides (Iñiguez-Moreno et al., 2018).

Indwelling medical instruments like intratracheal, urinary central vein and peritoneal dialysis nephrostomes have become essential in clinical settings and patients management. These life-saving devices are prone to microbial attacks, especially biofilms. The biofilms on these devices are life-threatening to already vulnerable patients. Biofilm related infections have become a major cause of morbidity and mortality among hospitalized patients (Ledwoch et al., 2018). Bacterial biofilms have caused more than half of immunocompromised patients to develop bacterial infections. As per U.S. National institute of Health, these infections are as high as 80% due to these biofilms (Bridier et al., 2015, Mazzoli, 2010). Due to huge scale of this issue, multiple chemicals, physical and biological method have been tested. Each method has its own advantages and limitations. Huge investments are underway to target the biofilms on multiple surfaces especially on medical devices. These methods have proved effective alone while some require the assistance of other methods.

Physical Strategies

Biofilms are controlled by a multitude of including physical, chemical, and biological methods. Physical methods have been largely employed to control biofilm formation on multiple surfaces. These methods include ultrasound treatment, photodynamic therapy, electric field, magnetic field, plasma, electrolyzed water treatment, and ionizing radiations. This chapter will discus multiple aspects of each of the methods with specific examples (Table 1).

Physical Debridement

Biofilm infections in wounds make them hard to treat with simple antibiotics. These resistant microbes of biofilms sometimes need to be extracted from their roots. Wounds attached to biofilms can penetrate other tissues, which is a clinical menace. Physical debridement of biofilms is routinely utilized in wound care. This method efficiently removes biofilm beds from wounds, ultimately reducing the viable count of microbes and removing dead tissue (Malone et al., 2017). Some biofilms debridements are weekly performed like in critical limb ischemia patients. A kit and device method was filed in 2005 for biofilm debridement. Although sharp methods are painful, various less painful antibiofilm strategies are required (Ammons, 2010).

Table 1. Different physical methods commonly employed to carry out antibiofilm activity

Physical Method	Biofilm producing bacterial species	Additional Treatment (if any)	Reference
Ultrasound Therapy	*Staphylococcus aureus*	Chlorine Dioxide	(Yu et al., 2021)
	Staphylococcus epidermidis RP62A	Vancomycin	(He et al., 2011)
	Listeria monocytogenes	Benzalkonium Chloride	(Torlak et al., 2013)
	Pseudomonas diminuta	-	(Zips et al., 1990)
	Escherichia coli	Antibiotics	(Carmen et al., 2005)
	K. pneumonia	-	(Islam et al., 2017)
Photodynamic Therapy	*Escherichia coli, Staphylococcus aureus,* and methicillin-resistant *S. aureus*	Chemotherapy	(Chen et al., 2019)
	S. aureus	Quorum Sensing Inhibition	(Sun et al., 2019)
	Staphylococcus aureus and *Pseudomonas aeruginosa*	-	(Vassena et al., 2014)
	Staphylococcus aureus	Riboflavin	(Banerjee et al., 2020)
Electric Field	*MRSA, E. coli, P. aeruginosa*	-	(Berthelot et al., 2017)
	S. aureus, S. epidermidis, P. aeruginosa	Antimicrobial Compounds	(Del Pozo, Rouse, Mandrekar, Sampedro et al., 2009)
	E. coli	-	(Kim et al., 2015)
	S. epidermidis, S. aureus, E.coli, C. albicans	-	(Ruiz-Ruigomez et al., 2016)
	S. epidermidis	Orthopaedic Implants	(Pickering et al., 2003)
	E coli	Gentamicin and with Oxytetracycline	(Caubet et al., 2004)
	E. faecalis	Chlorhexidine	(Froughreyhani et al., 2018)
	A. baumannii, P. aeruginosa	Wireless Electroceutical Dressing (WED)	(Barki et al., 2019)

Physical Method	Biofilm producing bacterial species	Additional Treatment (if any)	Reference
Plasma	*E. coli, MRSA, S. epidermidis*	Microwave Induced Plasma	(Lee et al., 2009)
	C. albicans, S. aureus P. aeruginosa	Pretreatment with Ascorbic acid	(Taghizadeh et al., 2015)
	Pseudomonas aeruginosa, Burkholderia cenocepacia and Escherichia coli		(Ermolaeva et al., 2011)
Magnetic field	*Staphylococcus aureus* and *Escherichia coli*	Antibiotic Gentamicin (Gen), Tannic acid, and Silver nanoparticles (AgNPs)	(Wang, Wu, et al., 018)
	Pseudomonas aeruginosa	Gentamicin	(Benson et al., 1994)
	S. aureus		(Wang, Deng, et al., 2018)
Ionizing radiations	*Salmonella typhimurium*	Aqueous Chlorine dioxide	(Park et al., 2018)
	Salmonella enteritidis	-	(Zhang et al., 2021)
	Listeria monocytogenes	-	(Niemira, 2010)
	L. monocytogenes, E. coli	-	(Adhikari et al., 2015)

Ultrasound Treatment

Ultrasound (US) treatment create shear forces that can damage the biofilm structure owing to the bubble and fluid flow process during the sonication phenomenon (Bigelow, 2008, Oulahal-Lagsir et al., 2000, Yu et al., 2020). In liquid, ultrasonic waves create expansion and compression zones, this pressure change leads to the cavitation process and stimulate small bubble nuclei in liquid media. This bubble undergoes various processes *i.e.,* oscillation, growth, and contraction. Moreover, disruption of this cavitating bubble generates high shear forces, localized high pressure, and temperature followed by accelerated shock waves that have the capability to disturb biofilms from a surface (Verhaagen et al., 2016). Processes of acoustic streaming, micro streamers, and microstreaming, are some of the major contributors to biofilm removal from the various surfaces (Nowak et al., 2015). Cell permeability and reduced mass transfer have been reported by the acoustic streaming process, which is generated by the momentum transfer from the acoustic wave to the liquid (Ananta et al., 2005).

Antibiofilm Activity of Ultrasound Therapy

Different studies have been carried out under different ultrasonic conditions to attain antibiofilm activity. High-intensity ultrasound process has higher chances of biofilm disruption and cell dislodgment rate as compared to low intensity. High-intensity US treatment has a greater chance of creating enough mechanical energy to destroy biofilm integral structure and attain a high level of cell detachment rate. In comparison to that, low-intensity ultrasound works by stimulating bacterial metabolism, create more resistance and adherence of biofilm cells (Erriu et al., 2014). Different power levels can significantly remove biofilms at a different rate. Moreover, medium power ultrasound can remove more biofilms as compared to low power rate (Vyas et al., 2016). High cavitation intensity (>10 W/cm^2) is required to detach biofilm cells (Erriu et al., 2014). Low frequency (35 kHz) US treatment inactivate more than 90% of *Listeria monocytogenes* in the biofilm formed on polystyrene surface (Torlak and Sert, 2013). Moreover, ultrasound methods ascended the efficacy of antibiotics against biofilms of different bacterial species (*E.coli, Acinetobacter baumannii, Staphylococcus aureus* biofilms) (Liu et al., 2016, Peterson et al.,

2000, Wang, Wen, et al., 2018). Researchers have also utilized ultrasound and enzyme synergistic treatment for removing biofilms, where enzymes have considerably enhanced the removal efficacy of biofilms by the destruction of EPS matrix (Oulahal-Lagsir et al., 2003, Oulahal et al., 2007).

In-vitro ultrasonication can significantly reduce the biofilms of *E. coli* and *Pseudomonas aeruginosa* by increasing the transport of antibiotics (gentamicin) across biofilm, thereby, enhancing the antimicrobial activity of biofilm (Carmen et al., 2004). Implants *E. coli* biofilms can be controlled by using low-frequency pulsed ultrasound in conjunction with gentamicin, a similar trend was not found associated with biofilms of *Pseudomonas aeruginosa* (Carmen et al., 2005, Rediske et al., 1999, Rediske et al., 2000). A combination of ultrasound at frequencies of 28-48 kHz with gentamicin can reduce the viability of *E. coli* biofilms by more than 50% (Ensing et al., 2005).

Photodynamic Therapy

Photodynamic therapy utilizes different components including a chemical known as photosensitizer (PS), light, and oxygen. The excited state of PS forms a triplet excited state, which transfers energy to biological molecules or molecular oxygen depending on the nature of a reaction. These reactive species could be lipid-derived radicals or hydroxyl radicals if the biological molecule was lipid in nature (Figure 1). These charged radicals can produce reactive oxygen species (ROS) like hydrogen peroxide, which are powerful antimicrobial and cytotoxic agents (Owen et al., 2013) causing cellular damage and even cell death (Simon et al., 2000). In another reaction, singlet-state oxygen can be generated during PS-based triple excitation state, which can cause oxidation of biological molecules leading to cell damage and death (Bragado et al., 2007). These radicals can combine or react with other biomolecules and oxygen to yield hydrogen peroxide, causing lipid peroxidation or leading to the production of reactive oxygen species that can cause cellular damage and cell death (Wainwright, 1998).

A synergistic system having antibiofilm activity has been examined using QS inhibition and antibiotics and photodynamic therapy based on hollow carbon nitride spheres (HCNSs). 48 hours of incubation controlled biofilm formation of *S. aureus* by 40%and this system when checked *in vivo*, changed virulence factor production and surface hydrophobicity, reducing biofilm formation and attachment. Drug resistance was decreased after using this system (Sun et al., 2019). Another nano-antibacterial platform was constructed by using indocyanine green (ICG) photosensitizers and silver nanoparticles (AgNPs) onto molybdenum disulfide (MoS_2) nanosheets surface. This platform performed triple-action antibacterial activity using photothermal/ photodynamic/chemo activities with 808 nm laser irradiation. It effectively controlled the deep penetration of *S. aureus* biofilms using *in vitro* conditions.

MoS_2/ICG/Ag system controlled *S. aureus* biofilms-infected wound *in vivo* with minimal toxicity (Li et al., 2022). APDT with RLP068/Cl, a novel photosensitizer reduced MRSA and *P. aeruginosa* biofilms on prosthetics surfaces and reduced viable cell count in biofilms (Vassena et al., 2014).

Electrical Field Treatment

Electric fields and electric current can create tremendous changes in cell activity, biological membranes, metabolic processes, and various other cellular activities of both types of cells (eukaryotic and prokaryotic) (Ravikumar et al., 2019). Repulsive electrostatic forces tend to generate a repulsive effect on the adhesion of microbes to the surface and electric forces overall can mimic this repulsive effect which can help in the detachment of biofilms from the surface, a phenomenon termed as 'electricidal effect' (Jucker et al., 1996, Rijnaarts et al., 1993). The 'bioelectric effect' works on the principle that electric current enhances the performance of antimicrobial agents and helps in controlling microbial growth (Van Der Borden et al., 2004)

The duration and intensity of electric current determine the detachment rate of cells from the biofilm surface. Low-intensity current (4-ampere) with 4-hour current provided sporadically per day can decrease the biofilms of *S. aureus, S. epidermidis,* and *P. aeruginosa* on a variety of surfaces (Teflon or titanium discs) (Schmidt-Malan et al., 2015). However, the short duration of electric current bursts cannot remove biofilms from the surface. Electrical current has been proved effective in eradicating biofilms ever from short term exposure of current (Berthelot and Neethirajan, 2017). However, prolonged exposure can drastically plummet (~6 log_{10} CFU/ml) the viability of *S. aureus, Staphylococcus epidermidis, and Pseudomonas aeruginosa* biofilms, even at low-intensity electric current (20-, 200-, and 2,000-microampere) (Del Pozo, Rouse, Mandrekar, Steckelberg, et al., 2009). The pulsed electric field has been demonstrated to impart irreversible damage to biofilm integral structure (Khan et al., 2016). A strong electric field with a low current intensity enhances the biocidic effect against *P. aeruginosa*. It was demonstrated that the presence of a strong electric field with a low current density increases the industrial effectiveness of a biocide against *P. aeruginosa*. Previously, long exposure to the electric field has proved effective in controlling biofilms without the assistance of antibiotics (Sadekuzzaman et al., 2015).

Electrolyzed Water

The electrolyzed water (EW) method is a time-tested and clinically certified method. It offers multiple benefits like being safe, non-corrosive, cost-effective, environment friendly, and easier to use. It is usually prepared by a cell having positive and negatively charged electrodes, separated by a membrane, through which diluted saltwater passes. Two types of water are usually generated after the current passes through the electrodes. On the anode, oxidizing water with a pH less than 2.5 is normally produced while on the cathode reducing or alkaline water is produced. When electrodes are subjected to a direct current voltage current, 2 types of water with different properties are generated. The anode generates acidic or oxidizing water ($pH < 2.5$), and the cathode produces alkaline or reducing water (Park et al., 2002, Rahman et al., 2016). Electrolyzed oxidizing water (EOW) is a popular disinfectant used in the food industry (Hricova et al., 2008, Huang et al., 2008, Izumi, 1999, Koseki et al., 2001) and in wound treatment (Yahagi et al., 2000). EOW possesses strong antimicrobial activity owing to the presence of hypochlorous acid (HOCl) in the solution. It can destroy biofilms of *L. monocytogenes* on a hard surface like stainless steel by 4.3 to 5.2 log CFU per coupon by only exposing biofilm coated surface for 30-120 seconds (Ayebah et al., 2005). Furthermore, alkaline water has a potent antibiofilm activity by removing glucans in the artificial biofilms of *S. mutans* (Gyo, 2009). It has been suggested that alkaline ion water (AW) has a strong ability to remove a remarkable amount of a biofilm by disintegrating glucans in the artificial biofilm of *S. mutans* (Gyo, 2009). Additionally, basic EW and acidic EW have bactericidal activity against *S aureus* biofilm that is comparable to 2% NaOH and 2% HCl, respectively (Sun et al., 2012). Extracellular DNA (eDNA) is the key factor in biofilm formation. *Vibrio parahaemolyticus* biofilms can be broken down by treating them with DNase I and acidic electrolytic water (AEW). Here DNase I targeted eDNA and EPS producing genes and AEW wiped out the structure of biofilm and prevent them from regrowing by 12h. Hence, AEW is proved more effective than enzyme treatment for controlling biofilms (Li et al., 2020).

Plasma

Plasma is the fourth state of matter after solid, liquid, and gas. Plasma is fully or partially ionized gas that is composed of ions, photons, ultraviolet rays, free electrons, and an element at a neutral charge (Nicholson et al., 1983). Plasma is classified into non-thermal plasma (NTP or cold plasma) and thermal plasma according to thermodynamics balance (Gilmore et al., 2018). In NTP, as electron gain, most of the coupled energy and other ions and neutral particles have almost no energy, so their temperature remains low whereas, thermal plasma has a high temperature reaching up to thousands of kelvin. (López et al., 2019). Nonthermal atmospheric pressure plasma devices have emerged which have been utilized to expose biofilm to plasma, directly or indirectly. In direct plasma exposure, the biofilm is exposed to high electric fields, UV and charged particles. Bacteria are inactivated through direct exposure inactivation of bacteria, whereas charge accumulation on the bacterial membrane and the resulting electrostatic force overcomes the tensile strength of the membrane, leading to rupture. Plasma also makes reactive oxygen-nitrogen species (RONS). These have the potential to damage cellular structures in any living cell (Baran et al., 2004). Depending on the different parameters, plasma will produce a variety of species i.e., radicals, ions, stable hydrogen peroxide (H_2O_2), ozone (O_3), nitric oxide (NO), and another unstable superoxide anion ($O_{2,}$), singlet oxygen, hydroxyl radical and more in the gas surrounding environment. These species can cause oxidative damage by oxidizing biological molecules, helping the fall of biofilm structure, decreasing biofilm attachment to the surface ultimately disintegrating biofilms (Gupta et al., 2018). Chemical bonds present in the membranes and biological molecules of biofilms can be broken down by the etching effect induced by charged molecules produced by plasma treatment. The etching effect ensures biofilm loses its shielding function against microbes and biofilms will fall from the solid support surface and detached floating cells will be targeted by the free radicals, ions, excited atoms which can penetrate the cellular surface (Gupta et al., 2018, Zhu et al., 2020).

He/O_2 plasma jet can inactivate the biofilm of *Porphyromonas gingivalis* up to a depth of 15 mm on sandblasted disc acid-etched (SLA) titanium discs as the duration of exposure increases, the biofilm will significantly get destroyed (Chen et al., 2021), whereas the cold plasma at room temperature penetrated to the depth of 25.5 μm for damaging *Enterococcus faecalis* biofilms. The duration of treatment was ~100 ns duration with the current of ~6 mA (Pei et al., 2012). Cold plasma treatment for a duration of 8-10 minutes successfully destroyed *Enterococcus faecalis* biofilms from the infected root canal. Scanning electron microscope (SEM) images confirmed the rupture of bacterial membrane followed by full annihilation of biofilm (Pan et al., 2013). A cold plasma jet *Enteroscoccus faecalis* biofilms killed 93% of *Enterococcus faecalis* biofilms on the surface of hydroxyapatite discs (Jiang et al., 2012). Argon plasma can decrease the thickness of *Pseudomonas aeruginosa* PA103 biofilm (Ermolaeva et al., 2011). Non-thermal plasma (He/O_2) can penetrate up to 40-80 mm in biofilms of *P. aeruginosa* PAO1 and control its further growth (Alkawareek et al., 2012).

Magnetic Field

A magnetic field is a region of space present in the surroundings of a magnet, changing the electric field in which magnetic forces are observed. The static magnetic field has been reported to increase the gentamicin activity against a *P. aeruginosa* biofilm. Magnetic fields between 5 and 20 G ensured maximal reduction of adherent bacterial cells on different surfaces (polystyrene glass, chronoflex, biomer (Ethicon, Somerville, NJ) (Benson et al., 1994). Bacterial biofilm on prosthetic implants can be controlled using controlled magnetic fields in conjunction with antibiotics that may help in controlling and clearing biofilms on surfaces (McLeod et al., 2010). Ciprofloxacin (Cipro), magnetic nanoparticles (MNP), and static magnetic fields successfully plummeted the *Pseudomonas aeruginosa* biofilms metabolism and biomass (Bandara et al., 2015). In addition to this combination, magnetic nanoparticles can assist antibiotics to penetrate deeper into a biofilm with the assistance of a magnetic field and create an artificial channel in the biofilm matrix, helping detachment of cells (Quan et al., 2021).

Ionizing Irradiations

Another physical strategy for anti-biofilm is ionizing irradiation. Ionizing radiations impart destructive changes in the microbial DNA ensuring the viable reduction of microbes (Howard-Flanders, 1959). This method is routinely applied for ensuring the sterilization of food products to meet export criteria (Figure 2) (Hayashi, 1991). Low doses and high doses of gamma irradiation have been routinely applied for the sterilization of medical devices (Govindaraj et al., 2015, Mubarak et al., 2019). Alpha radiations in conjunction with a monoclonal antibody (immunoglobulin G1 [IgG1]) have been shown to penetrate *Cryptococcus neoformans* biofilm and reduced its activity by 50%. This treatment has the potential to control biofilm formation on indwelling medical devices (Martinez et al., 2006). Three serovars of *S. enterica* depicted different sensitivities to different doses of gamma irradiations. *S. enterica* serovar biofilm was more sensitive as compared to its planktonic cell to ionizing radiations (Niemira et al., 2005). Ultraviolet (UV-C) is also an antibiofilm treatment (Motola et al., 2019). UV-C (Ultraviolet C) has been reported to control biofilms of *L. monocytogenes, S. typhimurium, Escherichia coli* O157: H7 (Adhikari et al., 2015). X-rays provides another exciting opportunity for controlling biofilms by providing deeper penetration into biofilms. X-rays can be used in combination with another disinfectant for augmenting their antibiofilm activity and are reported to reduce the biofilm of *Salmonella* strains on egg shells whereas in other studies, x-ray irradiation and Sodium hypochlorite synergistic effect significantly reduced the biofilms of *S. typhimurium* (Park, Jung, and Ha, 2018).

Conclusion

The arising resistance of microbes in biofilm against antibiotics is a threatening situation. New strategies are in need to be developed for mitigating the harms of biofilm in different industries, especially medical industry. Physical methods provide an efficient approach for controlling biofilms on different surfaces. Each method from the ultrasound method to low-temperature plasma has contributed significantly for controlling biofilms. But these techniques require more *in-vivo* studies before their wide-scale applications. The molecular phenomenon behind their detailed mechanism needs further research. These methods are still in the nascent phase and require further in depth research from methodology and cellular level perspective.

References

Adhikari, A., Syamaladevi, R. M., Killinger, K. and Sablani, S. S. (2015). Ultraviolet-C light inactivation of *Escherichia coli* O157: H7 and Listeria monocytogenes on organic fruit surfaces. *International Journal of Food Microbiology,* 210, 136-142.

Alkawareek, M. Y., Algwari, Q. T., Laverty, G., Gorman, S. P., Graham, W. G., O'Connell, D. and Gilmore, B. F. (2012). Eradication of *Pseudomonas aeruginosa* biofilms by atmospheric pressure non-thermal plasma.

Ammons, M. C. (2010). Anti-biofilm strategies and the need for innovations in wound care. *Recent patents on anti-infective drug discovery*, 5, 10-17.

Ananta, E., Volkert, M. and Knorr, D. (2005). Cellular injuries and storage stability of spray-dried *Lactobacillus rhamnosus* GG. *International dairy journal,* 15, 399-409.

Ayebah, B., Hung, Y. C. and Frank, J. F. (2005). Enhancing the bactericidal effect of electrolyzed water on *Listeria monocytogenes* biofilms formed on stainless steel. *Journal of Food Protection*, 68, 1375-1380.

Bandara, H., Nguyen, D., Mogarala, S., Osinski, M. and Smyth, H. (2015). Magnetic fields suppress *Pseudomonas aeruginosa* biofilms and enhance ciprofloxacin activity. *Biofouling,* 31, 443-457.

Banerjee, S., Ghosh, D., Vishakha, K., Das, S., Mondal, S. and Ganguli, A. (2020). Photodynamic antimicrobial chemotherapy (PACT) using riboflavin inhibits the mono and dual species biofilm produced by antibiotic resistant *Staphylococcus aureus* and *Escherichia coli. Photodiagnosis and Photodynamic Therapy*, 32, 102002.

Baran, C. P., Zeigler, M. M., Tridandapani, S. and Marsh, C. B. (2004). The role of ROS and RNS in regulating life and death of blood monocytes. *Current pharmaceutical design,* 10, 855-866.

Barki, K. G., Das, A., Dixith, S., Ghatak, P. D., Mathew-Steiner, S., Schwab, E., Khanna, S., Wozniak, D. J., Roy, S. and Sen, C. K. (2019). Electric field based dressing disrupts mixed-species bacterial biofilm infection and restores functional wound healing. *Annals of surgery,* 269, 756.

Benson, D. E., Grissom, C. B., Burns, G. L. and Mohammad, S. F. (1994). Magnetic field enhancement of antibiotic activity in biofilm forming *Pseudomonas aeruginosa.* ASAIO Journal (American Society for Artificial Internal Organs: 1992), 40, M371-376.

Berthelot, R. and Neethirajan, S. (2017). Harnessing electrical energy for anti-biofilm therapies: effects of current on cell morphology and motility. *Journal of Experimental Nanoscience,* 12, 197-207.

Bigelow, T. A. (2008). Ultrasound attenuation estimation using backscattered echoes from multiple sources. *The Journal of the Acoustical Society of America*, 124, 1367-1373.

Bragado, P., Armesilla, A., Silva, A. and Porras, A. (2007). Apoptosis by cisplatin requires p53 mediated p38α MAPK activation through ROS generation. *Apoptosis,* 12, 1733-1742.

Bridier, A., Sanchez-Vizuete, P., Guilbaud, M., Piard, J. C., Naitali, M. and Briandet, R. (2015). Biofilm-associated persistence of food-borne pathogens. *Food microbiology*, 45, 167-178.

Carmen, J. C., Roeder, B. L., Nelson, J. L., Ogilvie, R .L. R., Robison, R. A., Schaalje, G .B. and Pitt, W. G. (2005). Treatment of biofilm infections on implants with low-frequency ultrasound and antibiotics. *American journal of infection control*, 33, 78-82.

Carmen, J. C., Runyan, C. M., Robison, R. A., Nelson, J. L., Beckstead, B. L., Pitt, W. G. and Schaalje, G. B. (2004). Ultrasonic-enhanced gentamicin transport through colony biofilms of *Pseudomonas aeruginosa* and *Escherichia coli*. *Journal of Infection and Chemotherapy*, 10, 193-199.

Caubet, R., Pedarros-Caubet, F., Chu, M., Freye, E., de Belem Rodrigues, M., Moreau, J. and Ellison, W. (2004). A radio frequency electric current enhances antibiotic efficacy against bacterial biofilms. *Antimicrobial agents and chemotherapy*, 48, 4662-4664.

Chen, H., Yang, J., Sun, L., Zhang, H., Guo, Y., Qu, J., Jiang, W., Chen, W., Ji, J. and Yang, Y. W. (2019). Synergistic Chemotherapy and Photodynamic Therapy of Endophthalmitis Mediated by Zeolitic Imidazolate Framework-Based Drug Delivery Systems. *Small*, 15, 1903880.

Chen, Z. and Wirz, R. E. (2021). Cold Atmospheric Plasma (CAP) Technology and Applications. *Synthesis Lectures on Mechanical Engineering*, 6, i-191.

Costerton, J. W. (1999). Introduction to biofilm. *International journal of antimicrobial agents*, 11, 217-221.

Costerton, J. W., Lewandowski, Z., Caldwell, D. E., Korber, D. R. and Lappin-Scott, H.M. (1995). Microbial biofilms. *Annual review of microbiology*, 49, 711-745.

Del Pozo, J. L., Rouse, M. S., Mandrekar, J. N., Sampedro, M. F., Steckelberg, J. M. and Patel, R. (2009). Effect of electrical current on the activities of antimicrobial agents against *Pseudomonas aeruginosa*, *Staphylococcus aureus*, and *Staphylococcus epidermidis* biofilms. *Antimicrobial agents and chemotherapy*, 53, 35-40.

Del Pozo, J. L., Rouse, M. S., Mandrekar, J. N., Steckelberg, J. M. and Patel, R. (2009). The electricidal effect: reduction of Staphylococcus and Pseudomonas biofilms by prolonged exposure to low-intensity electrical current. *Antimicrobial agents and chemotherapy*, 53, 41-45.

Ensing, G., Roeder, B., Nelson, J. L. E., Van Horn, J., Van Der Mei, H., Busscher, H. and Pitt, W. (2005). Effect of pulsed ultrasound in combination with gentamicin on bacterial viability in biofilms on bone cements in vivo. *Journal of applied microbiology*, 99, 443-448.

Ermolaeva, S. A., Varfolomeev, A. F., Chernukha, M. Y., Yurov, D. S., Vasiliev, M. M., Kaminskaya, A. A., Moisenovich, M. M., Romanova, J. M., Murashev, A. N. and Selezneva, I. I. (2011). Bactericidal effects of non-thermal argon plasma in vitro, in biofilms and in the animal model of infected wounds. *Journal of medical microbiology*, 60, 75-83.

Erriu, M., Blus, C., Szmukler-Moncler, S., Buogo, S., Levi, R., Barbato, G., Madonnaripa, D., Denotti, G., Piras, V. and Orrù, G. (2014). Microbial biofilm modulation by ultrasound: current concepts and controversies. *Ultrasonics sonochemistry*, 21, 15-22.

Flemming, H. C. and Wingender, J. (2010). The biofilm matrix. *Nature reviews microbiology*, 8, 623-633.

Flemming, H. C., Wingender, J., Szewzyk, U., Steinberg, P., Rice, S. A. and Kjelleberg, S. (2016). Biofilms: an emergent form of bacterial life. *Nature Reviews Microbiology*, 14, 563-575.

Froughreyhani, M., Salemmilani, A., Mozafari, A. and Hosein-Soroush, M. (2018). Effect of electric currents on antibacterial effect of chlorhexidine against *Entrococcus faecalis* biofilm: An in vitro study. *Journal of clinical and experimental dentistry,* 10, e1223.

Gilmore, B. F., Flynn, P. B., O'Brien, S., Hickok, N., Freeman, T. and Bourke, P. (2018). Cold plasmas for biofilm control: opportunities and challenges. *Trends in biotechnology,* 36, 627-638.

Govindaraj, S. and Muthuraman, M. S. (2015). Systematic review on sterilization methods of implants and medical devices. I*nt J ChemTech Res,* 8, 897-911.

Gupta, T. T., Karki, S. B., Fournier, R. and Ayan, H. (2018). Mathematical modelling of the effects of plasma treatment on the diffusivity of biofilm. *Applied Sciences*, 8, 1729.

Gyo, M. (2009). Effects of alkali-ion Water on single species *Streptococcus mutans* biofilm. *Int Chin J Dent,* 9, 55-60.

Hayashi, T. 1991. "Comparative effectiveness of gamma-rays and electron beams in food irradiation." In *Food irradiation.*

He, N., Hu, J., Liu, H., Zhu, T., Huang, B., Wang, X., Wu, Y., Wang, W. and Qu, D. (2011). Enhancement of vancomycin activity against biofilms by using ultrasound-targeted microbubble destruction. *Antimicrobial agents and chemotherapy*, 55, 5331-5337.

Howard-Flanders, P. 1959. "Physical and chemical mechanisms in the injury of cells by ionizing radiations." In *Advances in biological and medical physics*. 553-603. Elsevier.

Hricova, D., Stephan, R. and Zweifel, C. (2008). Electrolyzed water and its application in the food industry. *Journal of food protection*, 71, 1934-1947.

Huang, Y. R., Hung, Y. C., Hsu, S. Y., Huang, Y. W. and Hwang, D. F. (2008). Application of electrolyzed water in the food industry. *Food control*, 19, 329-345.

Iñiguez-Moreno, M., Gutiérrez-Lomelí, M., Guerrero-Medina, P. J. and Avila-Novoa, M. G. (2018). Biofilm formation by *Staphylococcus aureus* and *Salmonella* spp. under mono and dual-species conditions and their sensitivity to cetrimonium bromide, peracetic acid and sodium hypochlorite. *brazilian journal of microbiology,* 49, 310-319.

Islam, M. A., Woon, C. W., Ethiraj, B., Cheng, C. K., Yousuf, A. and Khan, M. M. R. (2017). Ultrasound driven biofilm removal for stable power generation in microbial fuel cell. *Energy & Fuels,* 31, 968-976.

Izumi, H. (1999). Electrolyzed water as a disinfectant for fresh-cut vegetables. *Journal of Food Science,* 64, 536-539.

Jiang, C., Schaudinn, C., Jaramillo, D. E., Webster, P. and Costerton, J. W. (2012). In vitro antimicrobial effect of a cold plasma jet against *Enterococcus faecalis* biofilms. *International Scholarly Research Notices*, 2012,

Jucker, B. A., Harms, H. and Zehnder, A. (1996). Adhesion of the positively charged bacterium *Stenotrophomonas (Xanthomonas) maltophilia* 70401 to glass and Teflon. *Journal of bacteriology,* 178, 5472-5479.

Khan, M. S. I., Lee, E. J. and Kim, Y. J. (2016). A submerged dielectric barrier discharge plasma inactivation mechanism of biofilms produced by *Escherichia coli* O157: H7, *Cronobacter sakazakii*, and *Staphylococcus aureus*. *Scientific Reports*, 6, 1-11.

Kim, Y. W., Subramanian, S., Gerasopoulos, K., Ben-Yoav, H., Wu, H. C., Quan, D., Carter, K., Meyer, M. T., Bentley, W. E. and Ghodssi, R. (2015). Effect of electrical energy on the efficacy of biofilm treatment using the bioelectric effect. npj *Biofilms and Microbiomes*, 1, 1-8.

Koseki, S., Yoshida, K., Isobe, S. and Itoh, K. (2001). Decontamination of lettuce using acidic electrolyzed water. *Journal of food protection,* 64, 652-658.

Ledwoch, K., Dancer, S., Otter, J., Kerr, K., Roposte, D., Rushton, L., Weiser, R., Mahenthiralingam, E., Muir, D. and Maillard, J. Y. (2018). Beware biofilm! Dry biofilms containing bacterial pathogens on multiple healthcare surfaces; a multi-centre study. *Journal of Hospital Infection,* 100, e47-e56.

Lee, M. H., Park, B. J., Jin, S. C., Kim, D., Han, I., Kim, J., Hyun, S. O., Chung, K. H. and Park, J. C. (2009). Removal and sterilization of biofilms and planktonic bacteria by microwave-induced argon plasma at atmospheric pressure. *New Journal of Physics,* 11, 115022.

Li, H., Gong, M., Xiao, J., Hai, L., Luo, Y., He, L., Wang, Z., Deng, L. and He, D. (2022). Photothermally activated multifunctional MoS2 bactericidal nanoplatform for combined chemo/photothermal/photodynamic triple-mode therapy of bacterial and biofilm infections. *Chemical Engineering Journal*, 429, 132600.

Li, Y., Tan, L., Guo, L., Zhang, P., Malakar, P. K., Ahmed, F., Liu, H., Wang, J. J. and Zhao, Y. (2020). Acidic electrolyzed water more effectively breaks down mature *Vibrio parahaemolyticus* biofilm than DNase I. *Food Control,* 117, 107312.

Limoli, D. H., Jones, C. J. and Wozniak, D. J. (2015). Bacterial extracellular polysaccharides in biofilm formation and function. *Microbiology spectrum*, 3, 3.3. 29.

Liu, X., Yin, H., Weng, C. X. and Cai, Y. (2016). Low-frequency ultrasound enhances antimicrobial activity of colistin–vancomycin combination against pan-resistant biofilm of *Acinetobacter baumannii. Ultrasound in medicine & biology,* 42, 1968-1975.

López, M., Calvo, T., Prieto, M., Múgica-Vidal, R., Muro-Fraguas, I., Alba-Elías, F. and Alvarez-Ordóñez, A. (2019). A review on non-thermal atmospheric plasma for food preservation: Mode of action, determinants of effectiveness, and applications. *Frontiers in microbiology,* 10, 622.

Malone, M. and Swanson, T. (2017). Biofilm-based wound care: the importance of debridement in biofilm treatment strategies. *British journal of community nursing*, 22, S20-S25.

Martinez, L., Bryan, R., Apostolidis, C., Morgenstern, A., Casadevall, A. and Dadachova, E. (2006). Antibody-guided alpha radiation effectively damages fungal biofilms. *Antimicrobial agents and chemotherapy*, 50, 2132-2136.

Mazzoli, S. (2010). Biofilms in chronic bacterial prostatitis (NIH-II) and in prostatic calcifications. *FEMS Immunology & Medical Microbiology*, 59, 337-344.

McLeod, B. R. and Sandvik, E. L. (2010). A biofilm growth protocol and the design of a magnetic field exposure setup to be used in the study of magnetic fields as a means of controlling bacterial biofilms. Bioelectromagnetics: Journal of the Bioelectromagnetics Society, The Society for Physical Regulation in Biology and Medicine, *The European Bioelectromagnetics Association*, 31, 56-63.

Motola, M., Dworniczek, E., Satrapinskyy, L., Chodaczek, G., Grzesiak, J., Gregor, M., Plecenik, T., Nowicka, J. and Plesch, G. (2019). UV light-induced photocatalytic, antimicrobial, and antibiofilm performance of anodic TiO_2 nanotube layers prepared on titanium mesh and Ti sputtered on silicon. *Chemical Papers,* 73, 1163-1172.

Mubarak, M. T., Ozsahin, I. and Ozsahin, D. U. 2019. "Evaluation of sterilization methods for medical devices." *2019 Advances in Science and Engineering Technology International Conferences* (ASET).

Nicholson, D. R. and Nicholson, D. R. 1983. *Introduction to plasma theory.* Wiley New York.

Niemira, B. A. (2010). Irradiation sensitivity of planktonic and biofilm-associated Listeria monocytogenes and L. innocua as influenced by temperature of biofilm formation. *Food and Bioprocess Technology*, 3, 257-264.

Niemira, B. A. and Solomon, E. B. (2005). Sensitivity of planktonic and biofilm-associated Salmonella spp. to ionizing radiation. *Applied and Environmental Microbiology*, 71, 2732-2736.

Nowak, T., Cairós, C., Batyrshin, E. and Mettin, R. 2015. "Acoustic streaming and bubble translation at a cavitating ultrasonic horn." *AIP Conference Proceedings.*

Oulahal-Lagsir, N., Martial-Gros, A., Boistier, E., Blum, L. and Bonneau, M. (2000). The development of an ultrasonic apparatus for the non-invasive and repeatable removal of fouling in food processing equipment. *Letters in Applied Microbiology,* 30, 47-52.

Oulahal-Lagsir, N., Martial-Gros, A., Bonneau, M. and Blum, L. J. (2003). "Escherichia coli-milk" biofilm removal from stainless steel surfaces: synergism between ultrasonic waves and enzymes. *Biofueling,* 19, 159-168.

Oulahal, N., Martial-Gros, A., Bonneau, M. and Blum, L. (2007). Removal of meat biofilms from surfaces by ultrasounds combined with enzymes and/or a chelating agent. *Innovative Food Science & Emerging Technologies,* 8, 192-196.

Owen, J. A., Punt, J. and Stranford, S. A. 2013. *Kuby immunology.* WH Freeman New York, NY, USA:

Pan, J., Sun, K., Liang, Y., Sun, P., Yang, X., Wang, J., Zhang, J., Zhu, W., Fang, J. and Becker, K. H. (2013). Cold plasma therapy of a tooth root canal infected with Enterococcus faecalis biofilms in vitro. *Journal of endodontics,* 39, 105-110.

Park, H., Hung, Y. C. and Kim, C. (2002). Effectiveness of electrolyzed water as a sanitizer for treating different surfaces. *Journal of food protection*, 65, 1276-1280.

Park, S. Y., Jung, S. J. and Ha, S. D. (2018). Synergistic effects of combined X-ray and aqueous chlorine dioxide treatments against Salmonella Typhimurium biofilm on quail egg shells. *LWT*, 92, 54-60.

Pei, X., Lu, X., Liu, J., Liu, D., Yang, Y., Ostrikov, K., Chu, P. K. and Pan, Y. (2012). Inactivation of a 25.5 μm *Enterococcus faecalis* biofilm by a room-temperature, battery-operated, handheld air plasma jet. *Journal of Physics D: Applied Physics*, 45, 165205.

Peterson, R. V. and Pitt, W. G. (2000). The effect of frequency and power density on the ultrasonically-enhanced killing of biofilm-sequestered Escherichia coli. *Colloids and Surfaces B: Biointerfaces*, 17, 219-227.

Pickering, S., Bayston, R. and Scammell, B. (2003). Electromagnetic augmentation of antibiotic efficacy in infection of orthopaedic implants. The Journal of bone and joint surgery. *British volume,* 85, 588-593.

Quan, K., Zhang, Z., Ren, Y., Busscher, H. J., van der Mei, H. C. and Peterson, B. W. (2021). Possibilities and impossibilities of magnetic nanoparticle use in the control of infectious biofilms. *Journal of Materials Science & Technology*, 69, 69-78.

Rahman, S., Khan, I. and Oh, D. H. (2016). Electrolyzed water as a novel sanitizer in the food industry: current trends and future perspectives. *Comprehensive Reviews in Food Science and Food Safety*, 15, 471-490.

Ravikumar, K., Basu, B. and Dubey, A. K. (2019). Analysis of electrical analogue of a biological cell and its response to external electric field. Regenerative *Engineering and Translational Medicine,* 5, 10-21.

Rediske, A. M., Roeder, B. L., Brown, M. K., Nelson, J. L., Robison, R. L., Draper, D. O., Schaalje, G. B., Robison, R. A. and Pitt, W. G. (1999). Ultrasonic enhancement of antibiotic action on *Escherichia coli* biofilms: an in vivo model. *Antimicrobial Agents and Chemotherapy,* 43, 1211-1214.

Rediske, A. M., Roeder, B. L., Nelson, J. L., Robison, R. L., Schaalje, G. B., Robison, R. A. and Pitt, W. G. (2000). Pulsed ultrasound enhances the killing of *Escherichia coli* biofilms by aminoglycoside antibiotics in vivo. *Antimicrobial agents and chemotherapy,* 44, 771-772.

Rijnaarts, H. H., Norde, W., Bouwer, E. J., Lyklema, J. and Zehnder, A. J. (1993). Bacterial adhesion under static and dynamic conditions. *Applied and Environmental Microbiology,* 59, 3255-3265.

Ruiz-Ruigomez, M., Badiola, J., Schmidt-Malan, S. M., Greenwood-Quaintance, K., Karau, M. J., Brinkman, C. L., Mandrekar, J. N. and Patel, R. (2016). Direct electrical current reduces bacterial and yeast biofilm formation. *International journal of bacteriology,* 2016,

Sadekuzzaman, M., Yang, S., Mizan, M. and Ha, S. (2015). Current and recent advanced strategies for combating biofilms. *Comprehensive Reviews in Food Science and Food Safety*, 14, 491-509.

Schmidt-Malan, S. M., Karau, M. J., Cede, J., Greenwood-Quaintance, K. E., Brinkman, C. L., Mandrekar, J. N. and Patel, R. (2015). Antibiofilm activity of low-amperage continuous and intermittent direct electrical current. *Antimicrobial agents and chemotherapy,* 59, 4610-4615.

SHAGUFTA, N. Effect of γ-irradiation doses on the sensory and microbial quality of dates (Phoenix dactylifera). *BIOLOGICAL SOCIETY OF PAKISTAN*, 227.

Sharahi, J. Y., Azimi, T., Shariati, A., Safari, H., Tehrani, M. K. and Hashemi, A. (2019). Advanced strategies for combating bacterial biofilms. *Journal of cellular physiology,* 234, 14689-14708.

Simon, H. U., Haj-Yehia, A. and Levi-Schaffer, F. (2000). Role of reactive oxygen species (ROS) in apoptosis induction. *Apoptosis,* 5, 415-418.

Souza, C. d., Faria, Y. V., Sant'Anna, L. d. O., Viana, V. G., Seabra, S. H., Souza, M. C. d., Vieira, V. V., Hirata, R., Moreira, L. d. O. and Mattos-Guaraldi, A. L. d. (2015). Biofilm production by multiresistant *Corynebacterium striatum* associated with nosocomial outbreak. *Memorias do Instituto Oswaldo Cruz*, 110, 242-248.

Sun, J. L., Zhang, S. K., Chen, J. Y. and Han, B. Z. (2012). Efficacy of acidic and basic electrolyzed water in eradicating *Staphylococcus aureus* biofilm. *Canadian journal of microbiology,* 58, 448-454.

Sun, Y., Qin, H., Yan, Z., Zhao, C., Ren, J. and Qu, X. (2019). Combating biofilm associated infection in vivo: integration of quorum sensing inhibition and photodynamic treatment based on multidrug delivered hollow carbon nitride sphere. *Advanced Functional Materials*, 29, 1808222.

Taghizadeh, L., Brackman, G., Nikiforov, A., van der Mullen, J., Leys, C. and Coenye, T. (2015). Inactivation of biofilms using a low power atmospheric pressure argon plasma jet; the role of entrained nitrogen. *Plasma Processes and Polymers,* 12, 75-81.

Torlak, E. and Sert, D. (2013). Combined effect of benzalkonium chloride and ultrasound against *Listeria monocytogenes* biofilm on plastic surface. *Letters in applied microbiology*, 57, 220-226.

Trancassini, M., Iebba, V., Citerà, N., Tuccio, V., Magni, A., Varesi, P., De Biase, R. V., Totino, V., Santangelo, F. and Gagliardi, A. (2014). Outbreak of *Achromobacter xylosoxidans* in an Italian cystic fibrosis center: genome variability, biofilm production, antibiotic resistance, and motility in isolated strains. *Frontiers in microbiology,* 5, 138.

Van Der Borden, A. J., Van Der Werf, H., Van Der Mei, H. C. and Busscher, H. J. (2004). Electric current-induced detachment of *Staphylococcus epidermidis* biofilms from surgical stainless steel. *Applied and environmental microbiology,* 70, 6871-6874.

Vassena, C., Fenu, S., Giuliani, F., Fantetti, L., Roncucci, G., Simonutti, G., Romanò, C. L., De Francesco, R. and Drago, L. (2014). Photodynamic antibacterial and antibiofilm activity of RLP068/Cl against *Staphylococcus aureus* and *Pseudomonas aeruginosa* forming biofilms on prosthetic material. *International journal of antimicrobial agents,* 44, 47-55.

Verhaagen, B., Zanderink, T. and Rivas, D. F. (2016). Ultrasonic cleaning of 3D printed objects and Cleaning Challenge Devices. *Applied Acoustics*, 103, 172-181.

Vyas, N., Sammons, R., Addison, O., Dehghani, H. and Walmsley, A. (2016). A quantitative method to measure biofilm removal efficiency from complex biomaterial surfaces using SEM and image analysis. *Scientific reports,* 6, 1-10.

Wainwright, M. (1998). Photodynamic antimicrobial chemotherapy (PACT). The *Journal of antimicrobial chemotherapy,* 42, 13-28.

Wang, J., Wen, K., Liu, X., Weng, C. X., Wang, R. and Cai, Y. (2018). Multiple low frequency ultrasound enhances bactericidal activity of vancomycin against methicillin-resistant *Staphylococcus aureus* biofilms. *BioMed research* international, 2018,

Wang, X., Deng, A., Cao, W., Li, Q., Wang, L., Zhou, J., Hu, B. and Xing, X. (2018). Synthesis of chitosan/poly (ethylene glycol)-modified magnetic nanoparticles for antibiotic delivery and their enhanced anti-biofilm activity in the presence of magnetic field. *Journal of materials science,* 53, 6433-6449.

Wang, X., Wu, J., Li, P., Wang, L., Zhou, J., Zhang, G., Li, X., Hu, B. and Xing, X. (2018). Microenvironment-responsive magnetic nanocomposites based on silver nanoparticles/gentamicin for enhanced biofilm disruption by magnetic field. *ACS applied materials & interfaces*, 10, 34905-34915.

Watnick, P. and Kolter, R. (2000). Biofilm, city of microbes. *Journal of bacteriology,* 182, 2675-2679.

Yahagi, N., Kono, M., Kitahara, M., Ohmura, A., Sumita, O., Hashimoto, T., Hori, K., Ning-Juan, C., Woodson, P. and Kubota, S. (2000). Effect of electrolyzed water on wound healing. *Artificial Organs*, 24, 984-987.

Yu, H., Liu, Y., Li, L., Guo, Y., Xie, Y., Cheng, Y. and Yao, W. (2020). Ultrasound-involved emerging strategies for controlling foodborne microbial biofilms. *Trends in Food Science & Technology,* 96, 91-101.

Yu, H., Liu, Y., Yang, F., Xie, Y., Guo, Y., Cheng, Y. and Yao, W. (2021). Synergistic efficacy of high-intensity ultrasound and chlorine dioxide combination for *Staphylococcus aureus* biofilm control. *Food Control*, 122, 107822.

Yuan, L., Sadiq, F. A., Wang, N., Yang, Z. and He, G. (2020). Recent advances in understanding the control of disinfectant-resistant biofilms by hurdle technology in the food industry. *Critical Reviews in Food Science and Nutrition*, 1-16.

Zhang, H., Chen, S. P., Seck, H. L. and Zhou, W. (2021). Low-energy X-ray inactivation of *Salmonella Enteritidis* on shell eggs in mono-/co-culture biofilms with *Pseudomonas fluorescens*. *Food Control,* 123, 107742.

Zhu, Y., Li, C., Cui, H. and Lin, L. (2020). Feasibility of cold plasma for the control of biofilms in food industry. *Trends in Food Science & Technology,* 99, 142-151.

Zips, A., Schaule, G. and Flemming, H. (1990). Ultrasound as a means of detaching biofilms. *Biofouling,* 2, 323-333.

Chapter 5

Emerging Trends in Antibiofilm Approaches: A Review

Noor Muhammad[1] and Iram Liaqat[1*]
[1]Department of Zoology, Government College University, Lahore, Pakistan

Abstract

A biofilm results when various microorganisms form a microbial community. Most biofilms are complex in structure and can be diverse in species. The microbes communicate via cell signaling molecules (quorum sensing (QS)) mechanisms. Biofilms can be harmful, particularly in biological systems, as they are hard to eradicate and are resistant to anti-microbial compounds. Therefore, biofilms are a significant health-related threat across the world. Conventional methods used to combat biofilms, such as physical, chemical, antibiotics, antiseptics, etc. are often not effective and can cause environmental problems. So, various novel approaches are being proposed and adopted by the scientific community. Among these approaches, the use of nanoparticles (NPs) is a promising emerging system as even tough and complex biofilms can be eradicated because of their strong anti-microbial properties and mode of action. Bioactive compounds from natural sources, mostly plants, are already being used as anti-biofilm agents. Many compounds like halogenated furanones, 2-amminoimidazole alkaloids, and flavonoids have been isolated from plants and marine organisms that have high anti-biofilm properties. Because of their broad-spectrum activities and non-specific mechanism of action, anti-microbial peptides (AMPs) are also potential candidates. AMP-based anti-biofilm strategies like prophylactic and therapeutic strategies are proving helpful.

---

*Corresponding Author's Email: iramliaq@hotmail.com.

In: Understanding Antibiofilm Activity
Editor: Iram Liaqat
ISBN: 978-1-68507-927-7

The combination of NPs with these natural products is proving helpful in fighting biofilm-based infections. Similarly, the use of combinations of antimicrobial/antibiofilm agents in suitable proportions can help to eradicate the biofilms and cure the disorders caused by biofilms. This review provides information on the emerging trends and approaches for the eradication of biofilms and the treatment of medical disorders caused by biofilms.

Keywords: biofilms, antibiofilm, biofilm eradication, antibiofilm compounds, antimicrobial compounds, review

Introduction

Biofilms are highly organized communities of microbes associated with surfaces that include both biological and inert surfaces like catheters. These biofilms are responsible for many types of infectious diseases and numerous health problems such as urinary tract infections (UTIs), prostatitis, chronic wounds, and endocarditis (Worthington et al., 2012). Microbes of the same or different species may form colonies to form biofilms. According to some estimates, a dental plaque biofilm can contain more than 500 bacterial species. In the process of biofilm development, already attached cells produce polymers that help in adhesion, matrix formation and also alter the organism's phenotype in respect of gene transcription and growth rate. The produced polymers are polysaccharides and prevent the access of antibiotics to the biofilm. This phenomenon is also believed to protect microbial cells which are present in biofilm from stress, host immune cells, and also any adverse effect of the environment (Singh et al., 2017).

The extracellular polysaccharides are the physical barrier to antibiotics. Further, the microbes which are present in the biofilm have a different genetic and physical profile compared to when they are in the planktonic phase (Donlan, 2001). This physical profile helps microbial cells to develop resistance to antibiotics. Furthermore, microbes start producing extracellular polymeric substances (EPS) and upregulate the genes which produce efflux pumps, and these pumps purge out antibiotics. As biofilms are highly complex structures, all the subpopulations in the biofilm develop resistance against both antibiotics and the host defense system. Once the biofilm becomes mature, it is very hard to eradicate it from biological surfaces. As biofilms cannot be eradicated easily, they require very aggressive antibiotic therapy (Otto, 2013). In the case of biological devices, removal of the implant is often the only option (Campoccia et al., 2013). In this study, potential antibiofilm compounds will be explored in detail. In addition, the focus will be on the mechanism of action of important antibiofilm compounds. This latest literature on antibiofilm compounds will be summarized and commercially available antibiotics, increasing microbial disease resistance and the development of new antibiofilm compounds will be discussed.

Nanoparticles as Emerging Antibiofilm Agents

Nanoparticles and Nanomaterial

Nano is a Greek word that is used as a prefix for extremely small particles or material having a size in nanometers (nm). Nanoparticles (NPs) have more than 2 dimensions in the size range of 1-100 NM (ASTM International) (Alanazi et al., 2010; Tufail and Liaqat, 2021). NPs have unique physical chemical and electrical properties, in particular their large surface area to volume ratio and NPs are extensively being used in pharmaceuticals, biomedicine, and electrochemistry (Haruna et al., 2016). NPs can be synthesized from any organic or inorganic material. Inorganic material-based NPs have more potential therapeutic value as they have more stability and drug loading capacity (Alaqad and Saleh, 2016). NPs are also being used in drug delivery systems and various drug determinations by pharmaceutical companies (Huang et al., 2008).

Types of Nanoparticles

The NPs are of various types based on various sizes, shapes, compositions, and functions. NPs can be placed into three main classes based on Origin, Dimension, and structural organization as summarized in Figure 1 and Table 1 (Acharya et al., 2009; Abhilash, 2010; Hasan, 2015).

Techniques for the Delivery of Natural Products to Biofilms Using Nanoparticles

The antimicrobial activity of natural compounds and phytochemicals is well known. However, because of their low solubility in water and hydrophobicity, these compounds may not reach their target sites like biofilms and thus their antibiofilm activity is limited. To solve these problems nanocarriers and nano formulations are being utilized to deliver natural compounds to the target sites in the biofilm matrix to eradicate biofilms, inhibit biofilm formation, and lower the adherence capacity of microbial cells. The methodology to incorporate natural compounds with nanocarriers can be classified into three main groups i.e., surface functionalization, encapsulation, and nano emulsions.

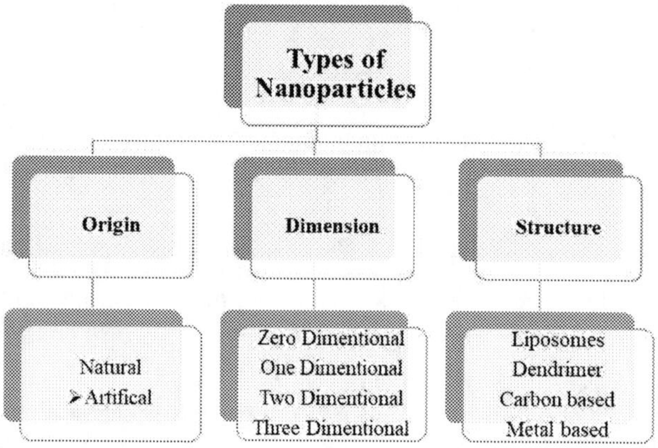

Figure 1. Types of nanoparticles.

Surface Functionalization

One or more natural products can be functionalized onto the NPs surface to enhance antibiofilm activity. Less soluble compounds can bind covalently or noncovalently with the NP's surface and their solvation property gets increased which ultimately increases the efficient targeting of a biofilm. It is known that NPs can penetrate the biofilm and this results in an increased local concentration of bioactive compounds resulting in a better and more effective antimicrobial effect (Barros et al., 2020).

Biogenically synthesized metal NPs (e.g., AgNPs) acquire a capping molecule during synthesis which helps their stability (Barros et al., 2018).. The synthesis of the carrier NPs takes place in the presence of an oil or any natural bioactive compound and a metal salt. At certain pH conditions, the NPs are synthesized by redox reaction and the bioactive compound oxidized and this oxidized active ingredient acts as capping agent for the NPs. The compounds can then reach their target site where they act as antimicrobial or antibiofilm agents (Menon et al., 2017; Sharma et al., 2019).

Table 1. Summary of the types of nanoparticles

Sr. No.	Nanoparticles	Antimicrobial/antibiofilm properties	References
1	Zinc Oxide	NPs get accumulated inside the cells; produce H_2O_2, the release of zinc ions disrupt the cell membrane	(Ren et al., 2009)
2	Titanium dioxide	Generation of reactive oxygen species results in the cell wall and cell membrane disruption	(Dastjerdi and Montazer 2010)
3	Copper oxide	Generation of reactive oxygen species which oxidize the lipids; free radical results in cell membrane disruption	(Qi et al., 2004)
4	Carbon nanotube	Production of reactive oxygen species disrupt the cell wall and proteins, and lipids get oxidized	(Chamundeeswari et al., 2010)
5	Chitosan	Increase membrane permeability which results in leaked out of intracellular material and deactivation of microbial enzymes	(Maness et al., 1999)
6	Gold	Generation of high electrostatic effect which disrupts the cell membrane	(Johnston and Hutchison, 2010)
7	Silver	The release of silver ions results in disruption of the cell membrane and DNA damage	(Kang et al., 2007)
8	Quaternary ammonia	Penetrate inside the cell and damage the cell wall	(Chen et al., 2020)

Nigella sativa essential oil coated onto gold nanoparticles (AuNPs) is effective against *Staphylococcus aureus* biofilms (Manju et al., 2016). Similarly, *Fucus vesiculosus* extract was applied on AuNPs and reported as effective against *Enterococcus faecalis* biofilm and was able to alter the bacterial surface hydrophobicity and exopolysaccharide production (Vijayakumar et al., 2017). Polyphenols extracted from *Usnea longissima* were capped on AuNPs. These capped NPs were able to inhibit the QS mechanism of *Streptococcus mutans* by decreasing the production of many virulence factors (Singh et al., 2015). Using a similar approach, copper NPs (CuNPs) were synthesized with *Eucalyptus globulus* extract and showed targeted antimicrobial effects (Ali et al. 2019). These types of NPs are also being used on the surfaces of medical devices having antiadhesive properties, thus preventing biofilm formation (Grumezescu et al., 2014). Another approach being used for the caping of NPs with natural bioactive compounds is the synthesis of NPs first and the adsorption of natural bioactive products on the surface of the formed NPs.

This method has a benefit as no change in the original active compound is taking place. Zinc NPs were synthesized and coated with natural products using such a technique and these not only inhibited the adherence of *Salmonella enterica* but also showed biofilm eradication properties (Stan et al., 2016). Nanoparticles as carriers of natural bioactive compounds are summarized in Table 2.

Encapsulation

The nanoencapsulation approach has one advantage over surface functionalization as this strategy enables sustained release of bioactive ingredients upon reaching the target site (Raza et al., 2019). It has been reported that nano capsules can be made to release the active compound at the target site only after a specific stimulus such as membrane matrix-like lower pH etc. (Elbialy et al., 2020; Avramović et al., 2020). Silica, lipids, polysaccharides, or synthetic polymers can be used to make nano capsules (Table 2) (Al-Qattan et al., 2018).

Mesoporous silica nanoparticles (MSNPs) have valuable properties as it is possible to tune their size, pore size, surface area, and sustained release of encapsulated compounds e.g., for hours or days (Rigby et al., 2008). A surfactant, such as cetrimonium bromide, can be used in the synthesis of MSNPs for the self-assembly of micelles which act as a template for hydrolysis or condensation of silica (Tang et al., 2012). After the displacement of micelles, the pores are free for drug encapsulation which can be done via adsorption or through wetness impregnation. In this process capillary action plays a role in the moving of small molecules to the inside of pores (van et al., 2011). Lipid NPs can be used for the encapsulation of hydrophobic bioactive compounds which can then be delivered to their target sites. The efficiency of encapsulation of compounds depends upon the lipid matrix, physical and chemical structure, and the polymorphic state of lipid (Müller et al., 2000). Solvent injection, ultrasonication, microemulsion dilution, and solvent evaporation are the most common lipid NPs preparation methods (Battaglia and Gallarate, 2012).

Table 2. Nanoparticles with various natural compounds showed antibiofilm activity were summarized

Natural Compound	Coated/Conjugated Nanoparticles	Tested Pathogens	Mode of Action	Antibiofilm effect	References
curcumin	SiO_2NPs	*S. aureus* and *Pseudomonas aeruginosa*	Cause ROS which results in oxidative stress	Showed bactericidal activity	(Mirzahosseinipour et al., 2020)
curcumin	Encapsulated NPs	*S. mutans*	Not known	Showed bactericidal activity	(Jahanizadeh et al., 2017)
curcumin	Nano hybrids	*S. mutans*	Not known	Showed bactericidal activity	(Maghsoudi et al., 2017)
Eugenol	Fe_3O_4 NPs	*P. aeruginosa* and *S. aureus*	Prevent cell adhesion	Show antibiofilm activity by killing bacterial cells	(Grumezescu et al., 2014)
Cinnamon Verum extract	Fe_3O_4 NPs	*Escherichia coli* and S. *aureus*	Prevent cell adhesion	Reduce hydrophobicity index up to 90% and decrease biofilm formation up to 77%	(Anghel et al., 2014)
Fucoidan	AuNPs	*Aeromonas hydrophila*	Hydrophobicity of bacterial membrane was changed	Reduced biofilm formation to 94%	(Vijayakumar et al., 2017)
curcumin	Curcumin Quantum dots	*E. coli* and *S. aureus*	Interfere with biofilm-forming proteins	Complete inhibition of biofilm formation	(Singh et al., 2017)
curcumin	Curcumin NPs	*S. aureus*	Cause ROS which results in oxidative stress	Not known	(Peng et al., 2019)
Curcumin	Graphene oxide NPs	*E. faecalis*	ROS generation and downregulation of genes that are responsible for biofilm formation	Inhibited biofilm formation at 250 µg/ml	(Ghorbanzadeh et al., 2020)
Curcumin	AgNPs	*E. coli, P. aeruginosa, Bacilis subtilis* and *S. aureus*	Not known	Reduction in biofilm formation up to 85%	(Jaiswal and Mishra, 2018)

Natural Compound	Coated/Conjugated Nanoparticles	Tested Pathogens	Mode of Action	Antibiofilm effect	References
Eucalyptus leaf extract	Copper NPs	*P. aeruginosa, E. coli* and *S. aureus*	ROS generation which results in oxidative damage	Reduce biofilm formation up to 70%	(Ali et al., 2019)
Curcumin	Graphene NPs	*C. albicans*	Prevention of cell adhesion	Reduced the biofilm formation ton 90%	(Palmieri et al., 2018)
Curcumin	Graphene NPs	*Prevotella intermedia*	ROS generation and downregulation of genes that are responsible for biofilm formation	Reduction biofilm formation to 76% and also showed bactericidal activity at 100µg/ml	(Pourhajibagher et al., 2019)
Curcumin	SiO_2 NPs	*P. putida*	Target proteins	Inhibited the biofilm formation to 50%	(Barros et al., 2020)
Usnic acid	ZnONPs	*S. enterica*	Prevention of cellular adhesion	Inhibited cellular adhesion to 80%	(Stan et al., 2016)
Carvacrol	Encapsulation in nanotubes	*A. hydrophila*	Inhibited the microbial cell adhesion	Not known	(Hendessi et al., 2016)
Propolis	Chitosan NPs	*Staphylococcus epidermis*	Downregulation of genes responsible for biofilm formation	Cause microbial cell mortality	(Ong et al., 2019)
Propolis	Chitosan NPs	*E .faecalis*	Downregulation of genes responsible for biofilm formation	Reduce biofilm formation by 75%	(Ong et al., 2017)
Propolis	Chitosan NPs	*E. faecalis*	Not known	Reduction of biofilm formation by 84%	(Carpio-Perochena et al., 2017)
Propolis	Polymeric NPs	*P. aeruginosa, S. aureus*	Not known	Inhibition of biofilm formation of up to 100%	(de et al., 2020)
Propolis	nanohydroxyapatite	*S. aureus*	Not known	Inhibition of cell adhesion	(Grenho et al., 2015)
Propolis	Fe_3O_4NPs	Methicillin resistant *S. aureus* (MRSA)	Prevention of cell adhesion	Inhibition of cell adhesion	(El-Guendouz et al., 2016)

Table 2. (Continued)

Garlic extract	Sol-gel NPs	MRSA	Not known	Inhibition of cell adhesion	(Girish et al., 2019)
Cinnamaldehyde	AuNPs	MRSA, MSSA, *E. coli* and *P. aeruginosa*	The cell wall was disrupted	Reduction of biofilm formation up to 97%	(Ramasamy et al., 2017b)
N. sativa essential oil	AuNPs	*S. aureus*	Hydrophobicity of bacterial membrane was changed	*S. aureus* hydrophobicity index was reduced to 78%	(Manju et al., 2016)
curcumin	Curcumin NPs	*E. faecalis*	Cause ROS which results in oxidative stress	Showed bactericidal activity	(Pourhajibagher et al., 2020)
curcumin	Curcumin NPs	*P. aeruginosa*	Downregulation of genes	Showed bactericidal activity	(Shariati et al., 2019)
Usnea longissimi extract	AuNPs	*S. mutans* and *Chromobacterium violaceum*	Inhibition of important enzymes like ATPase	Reduction of biofilm formation up to 90%	(B.N. Singh et al., 2015)
Clavanin A	SiO_2/Fe_3O4 NPs	*E. coli, P. aeruginosa, S. aureus* and *Klebsiella pneumonia*	The cell membrane was disrupted and also inhibited the cell adhesion	Decrease cellular adhesion up to 88%	(Ribeiro et al., 2018a, 2018b)
Cinnamaldehyde	SiO_2/AuNPs	MRSA, MSSA, *E. coli* and *P. aeruginosa*	The cell wall was disrupted	Prevent biofilm formation up to 100% and also showed biofilm eradication	(Ramasamy et al., 2017a)

Lipid nanovesicles have a natural affinity for essential oils because they contain hydrophilic polar moieties which faces the solvent, while hydrophobic molecules inside encapsulate the essential oil or bioactive compound (Shirodkar et al., 2019). Using this approach tea tree oil was encapsulated within a lipid nano carrier and delivered to the target site. It was effective against *Candida biofilms* (Souza et al., 2017). The mature biofilm of methicillin-resistant *S. aureus* (MRSA) bacterium was eradicated using thyme oil delivered using an encapsulation approach (Perez et al., 2019).

Nano-Emulsions

Nano-emulsions may be useful especially when the item which is to be delivered is an oil. Spontaneous emulsification or spontaneous emulsification using ultrasonication to emulsify oils into nanodroplets can be used to produce oil droplets of nanometer size (Lou et al., 2017; Rossi et al., 2017). This strategy has been reported many times in nano-based antibiofilm studies because it dispenses with the use of inorganic and nonmetallic carriers. Furthermore, this approach requires only one amphiphilic molecule which acts as a surfactant, but can have a co-surfactant which imparts more stability to the colloidal system (Kaur and Mehta, 2017; Ge et al., 2019). Essential oils are volatile with high reduction potential which means rapid oxidation, and they are sparingly soluble in water or an aqueous system. The delivery of essential oils as nano emulsions of nanodroplets overcomes this issue (Pavoni et al., 2020). Most of the nano antibiofilm literature reports the use of tween 80 as one of the important components for the preparations of nano emulsions and delivery of essential oils. Linaool, *Thymus daenensis* oil and *Cymbopogon flexuosus* oil have been successfully delivered using this approach which tends to enhance the antibiofilm activity of these oils as well (Table 2) (Prakash et al., 2019; Moghimi et al., 2018; da et al., 2018).

Antimicrobial/Antibiofilm Action of Nanoparticles

Nanoparticles function by penetrating the biofilm and by directly killing microbial cells (Tufail et al., 2022). The NPs enter into the bacterial cell wall and cause the release of ions or may remain deposited in the cell (Hoseinnejad et al., 2018). Due to their small size, they can cross the peptidoglycans layer of the cell wall and cell membrane (Lara et al., 2010). They damage the cellular membrane, disturb the cellular metabolism, produce reactive oxygen species (ROS) to cause oxidative stress, and inhibit transcription (Figure 2). The ions released by these NPs interact with various molecules found in the associated proteins of the cell wall and cell membrane (Hindi et al., 2009).

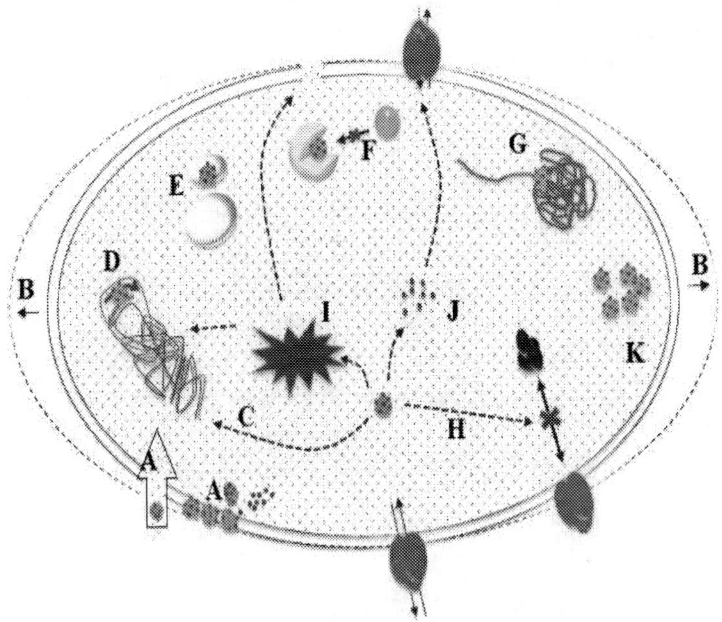

Figure 2. Antimicrobial action of nanoparticles (NPs). Adapted with permission from (Tufail and Liaqat, 2021). **A**. Nano silver alpha attaches to the cell membrane creating pores to cause cellular leakage, **B**. distortion of cellular morphology; **C**. breaking dsDNA; **D**. inhibition of DNA replication; **E**. interaction with 30S ribosome; **F**. inactivation of vital enzymes; **G**. denaturing proteins; **H**. modulation of cellular signaling; **I**. generating ROS, which acts on DNA and cell membranes; **J**. release of silver ions which affect the normal functioning of membrane protein; and **K**. accumulation inside the cell in lethal concentrations.

Furthermore, the contact of cationic ions with the bacterial cell membrane (negatively charged), makes the membrane porous and thus causes the release of intracellular content. This creates an ion imbalance in the cell and thus causes the release of ions into the cytoplasm via the cell membrane. Ions, thus interact with intracellular components resulting in cell death (Dakal et al., 2016). Many studies also have reported nanoparticle ions as inhibitors of proteins and enzyme activity that is involved in ATP production. They also inhibit respiratory enzymes which lead to ROS production, DNA and RNA damage, destabilize and disrupt the outer membrane (Blecher et al., 2011; Lara et al., 2010). The antibacterial effects of NPs are via inhibiting the synthesis of cell walls, nucleic acids, and proteins due to their interaction with the 30S ribosomal subunit (Shahverdi et al., 2007). Significant antibacterial potential of NPs against the resistant microbes is thus because of their ability to damage the bacterial cell in multiple ways, for example by disrupting intracellular pathways (Cui et al., 2013), damaging DNA and proteins (Li et al., 2010), and generating reactive oxygen within the cell (Kaweeteerawat et al., 2017).

Antimicrobial Peptides as Potential Antibiofilm Agents

Antimicrobial peptides also have been reported to have antibiofilm activity (AMPs). AMPs have been reported from different natural sources across all kingdoms (Wang, 2020). These include prokaryotes i.e., bacteria as well as eukaryotes i.e., animals including invertebrates, plants, fungi, and protozoans (Bahar and Ren, 2013). At present, there are over 5000 known types of both natural and synthetic AMPs (Zhao et al., 2013; Sadredinamin et al., 2016). They play an important role in the defense mechanisms of innate immunity and thus possess antimicrobial and antibiofilm potential (Figure 3).

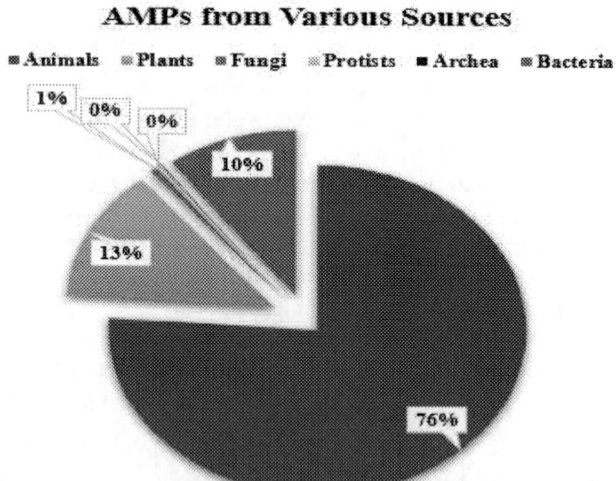

Figure 3. Various sources of antimicrobial peptides. Animals are the largest producers of antimicrobial peptides followed by plants and bacteria. Numbers obtained from http://aps.unmc.edu/AP/.

Many studies have already reported that AMPs inhibit biofilm formation at a lower concentration than the concentration required to kill microbial cells in planktonic mode. *S. aureus* and *P. aeruginosa* biofilms were inhibited by the synthetic AMP NA-CATH: ATRA1-ATRA1 and the natural AMP LL-37. It has been reported that these AMPs interfere with the gene which is responsible for biofilm development (Overhage et al., 2008). (Dean et al., 2011). LL-37 has also shown antiadhesion properties up to 62% against *S. aureus* and *P. aeruginosa* (Dean et al., 2011). In another study, AMP 1037 has shown antibiofilm activities by interfering with gene expressions, reducing motility, and reducing swimming of bacterial cells of *P. aeruginosa* (De et al., 2012).

How Antimicrobial Peptides Act

Antimicrobial peptides act in two different ways as summarized in Figure 4.

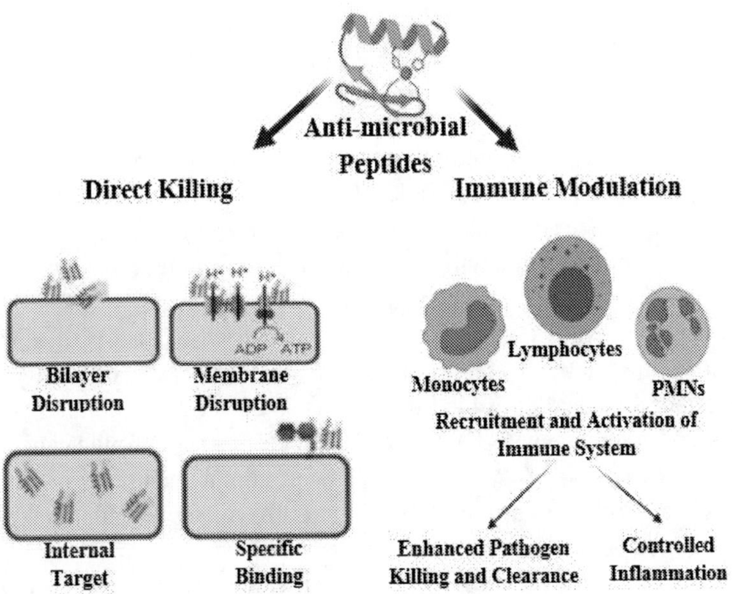

Figure 4. Mechanisms of action of antimicrobial peptides (Ulm et al. 2012). There are two main mechanisms of action i.e., Direct killing of cells and immune modulation.

Direct Killing: The Membrane-Targeting Mechanism

AMPs can show antimicrobial/antibiofilm potential with two types of membrane targeting mechanisms. The first is receptor-mediated and the other is non-receptor-mediated.

Receptor-Mediated Mechanisms

This pathway includes AMPs which are produced by bacteria and they are active in vitro in nanomolar concentrations, for example nisin (Shai, 2002; Shai and Oren, 2001). Nisin peptides consist of two domains, the first has a high affinity to the lipid II molecule in a membrane that is attached with the precursor of the cell wall. The second is the pore-forming domain which is membrane-embedded. Mesentericin can also be considered an example of receptor-mediated membrane-targeting AMPs (Shai and Oren, 2001; Kumar et al., 2018; Corrêa et al., 2019).

Non-Receptor-Mediated Mechanisms

Yeaman and Yount (2003) report that AMPs play their role by interacting with the components of the membrane. For example, the outer surface of Gram-positive bacteria and Gram-negative bacteria have teichoic & lipopolysaccharides, both having a net negative charge on the surface which allows electrostatic attraction with cationic AMPs.

Kumar et al. (2018) report that the mechanism of action of AMPs is different in the bacterial membrane and in the membrane of multicellular animals. The outer leaflet of the lipid bilayer in bacterial membranes is made of lipids that contain negatively charged head groups for example cardiolipin.

Guilhelmelli et al. (2013) report that, in animal membranes, the outer leaflet is made of zwitterionic phospholipids for example sphingomyelin & other neutral components like cholesterol. Guilhelmelli et al. (2013) and Zasloff (2002) report that lipids with negatively charged head groups are present in the inner leaflet which faces the cytoplasm in animal membranes. Epand et al. (2016) Andersson et al. (2016) report that the AMPs have stronger electrostatic interactions with the outer leaflet of the bacterial membranes than with animal membranes. After the electrostatic & hydrophobic interactions, the AMPs gather at the surface and, on reaching a certain concentration, they start to self-assemble on the bacterial membrane. At this stage, various models describe the action of AMPs. These models are divided into two categories:

Transmembrane Pore

The transmembrane pore models are subdivided into two categories. Barrel-stave pore and toroidal pore models. These were explained below.

Barrel-Stave Pore

In the barrel stave model, Park and Hahm, (2005) reported that the AMPs are oriented parallel to the membrane at the start & then insert in a perpendicular direction in the lipid bilayer, this gives rise to lateral peptide-peptide interactions (Wimley, 2010). Antimicrobial peptides for example alamethicin show barrel stave channeling. Huang et al., (2011) reported another pardaxin as another example as are protegrins (Brogden, 2005).

Toroidal Pore Model

In the toroidal pore model, peptides are in the perpendicular direction in the lipid bilayer but there are no specific peptide-peptide interactions (Wimley, 2010). These peptides form pores between the peptides and phospholipid head group (Uematsu and Matsuzaki, 2000). This active lipid-peptide supramolecule is known as the "toroidal pore." The distinctive characteristic of this model compared with the barrel-stave pore model is the net arrangement of the bilayer. In the barrel-stave pore model, the hydrophobic and hydrophilic arrangement of the lipids is maintained and in the toroidal pore model the hydrophobic and hydrophilic arrangement of the bilayer is not maintained. This gives rise to alternate surfaces for the interaction with the lipid tail and the lipid head group. As the toroidal pore is transient and, after the disintegration, these peptides move towards the inner cytoplasmic leaflet where they enter the cytoplasm and strongly target the intracellular components. (Yeaman and Yount, 2003) report that the toroidal pore has a discrete size and exhibits ion selectivity. It has been reported that AMPs such as magainin 2, lacticin Q exhibit this model activity (Lee et al., 2016). Both pore-forming models (toroidal pore and barrel) lead to membrane depolarization and finally cell death.

The Carpet Model

In this model, AMPs act without the formation of specific pores (Lee et al., 2016) and orientate parallel to the lipid bilayer (Yeaman and Yount, 2003). When they attain a threshold concentration, they cover the surface of the membrane like a "carpet." As a result, membrane integrity is lost. The same happens in the detergent model in which the membrane eventually disintegrates by forming micelles. The carpet model does not show peptide-peptide interactions of the membrane-bound peptide monomers; it also does not require the peptide to insert into the hydrophobic core to form transmembrane channels or specific peptide structures.

Direct Killing: Non-Membrane Targeting Mechanisms of Action

The non-membrane targeting of AMPs classified into two groups

- Bacterial cell wall target
- Intracellular targets

Bacterial Cell Wall Targeting

Like antibiotics AMPs inhibit cell wall synthesis (Malanovic and Lohner, 2016). The antibiotics attach to the specific proteins which are used in the synthesis of components of the cell wall while AMPs interact with a variety of precursor molecules that are used for the synthesis of the cell wall. Highly conserved lipid II is one of the molecules which is a major target (Figure 4). AMPs, such as defensins, bind with the negatively charged pyrophosphate sugar precursor of the lipid II molecule (Münch and Sahl, 2015). This binding leads to the formation of pores & membrane destruction, (Malanovic and Lohner, 2016).

Intracellular Targets

At first it was considered that AMPs have no intracellular targets (Brogden, 2005). Now it is recognized that many AMPs have intracellular targets as they cause bacterial death without causing permeabilization of the membrane at their minimum concentration.

In this mechanism, the AMPs show interaction with the membrane of cytoplasm and then they enter the cell where they disrupt cellular processes like protein or nucleic acid synthesis & block enzymatic activity. As an example, buforin II is an AMPs derived from frogs that enters the bacterial membrane of *E. coli* without its permeabilization & binds with DNA & RNA (Patrzykat et al., 2002; Brogden, 2005).

Immune Modulation Mechanism

As well as the direct killing of microbes, AMPs activate immune cells. As a result there is both an increase in microbial death and control of inflammation (Hilchie et al., 2013; Mader and Hoskin, 2006). AMPs are produced from certain immune cells like neutrophils & macrophages, so are considered as some of the first molecules that interact with attacking microbes (Jenssen et al., 2006). Examples of these AMPs are LL-37 & β defensins which activate the immune cells by chemoattraction, like mast cells. The activation of another group of immune cells (leukocytes) has also been reported (Niyonsaba et al., 2002).

Antimicrobial Peptides: A Hope against Resistance

Antimicrobial resistance is an emerging global health issue. The incidence multidrug resistance (MDR) is on the rise. Overuse and over prescription of some drugs is one of the causes of MDR (Michael et al., 2014). The WHO predict that, by 2050, 10 million people might be infected with drug-resistant microbes annually, due to the intense use of antibiotics (de et al., 2016; Adedeji 2016). Many MDR bacterial diseases, such as infections by *S. aureus,* are spreading in hospitals affecting 20% of people worldwide (Malanovic and Lohner, 2016). As the resistance in microorganisms is occurring very fast, finding alternatives to existing drugs is vital (Koo and Seo, 2019). Antimicrobial peptides are also named host defense peptides because they are very effective against microbial infection (Rossi et al., 2014). They are emerging defenses against MDR pathogens due to their different modes of action (Mookherjee et al., 2020). All AMPs possess two common features. First, they all possess a positive charge and second, 50% of their structure composition is hydrophilic so they can contact the membrane directly (Andersson et al., 2016; Waghu et al., 2016). Some AMPs e.g., defensin, destroy bacteria by making a hole in their membrane causing lysis (Wang, 2015).

They are known to be effective against Gram-positive and negative bacteria by interacting with lipoteichoic acid and lipopolysaccharides present in the cell wall and membranes respectively (Yasir et al., 2018). Some AMPs act by interfering with components inside cells rather than membranes (Le et al., 2017). Antimicrobial peptides, due to their multiple target sites, are very effective and this property is also a reason that AMPs combat MDR organisms (LaRock and Nizet, 2015). Antimicrobial peptides are also used in combination with antibiotics to combat multidrug resistance (Ventola, 2015). Bacteriocin, obtained from bacteria is an example of an AMP that can become an alternative to current antibiotics against MDR, and works by interfering with the membrane of the target bacteria (Simons et al., 2020) As AMPs have many advantages over currently used antibiotics against which bacteria are developing resistance day by day, they are a new hope against resistance. (Mwangi et al., 2019, Haney et al., 2019).

Quorum Sensing Inhibitors as Anti-Biofilm Agents

The QS phenomenon plays a very important role in the formation and regulation of the biofilm. Targeting this microbial communication system can significantly hinder biofilm formation and solve biofilm-related problems. *Amomum tsaoko* fruit extract has been reported to have anti-QS activity. The extract has inhibited biofilm formation of *S. typhimurium, P. aeruginosa,* and *S. aureus* by 52%, 45.3%, and 47% respectively when applied at a concentration of 4 mg/ml. The study further reported that the extract of *tsaoko* contains tsaokoaryline as an active ingredient which affects gene expression and motility in microbial cells (Rahman et al., 2017). A dietary phenolic phytocompound from sprouting barley, Hordenine, has been reported to have anti-QS activity against *P. aeruginosa*. Biofilm formation was reduced by 30% and preformed biofilm reduced by 23% (Zhou et al., 2018). *Delftia tsuruhatensis* SJ01, a Gram-negative bacterium, produces a bioactive compound that has shown anti-QS activity and biofilm inhibition in *P. aeruginosa* (Singh et al., 2017). A bioactive compound isolated from *Amphipterygium adstringens* shows anti-QS activity against pathogenic *P. aeruginosa*. This compound has inhibited the biofilm of *P. aeruginosa* by up to 94% even at low doses (Castillo-Juárez et al., 2013).

Antibiofilm Enzymes

Enzymes are the catalysts naturally found in biological systems, they enhance chemical reactions without being utilized in the reaction (Shanmugam, 2009). All biological and cellular systems depend on these enzymes, and any minor molecular modification may have severe metabolic consequences (Meireles et al., 2016). The activity and specificity of enzymes may be compromised by temperature, pH, presence or absence of activators, substrate, co-factors, and inhibitors (Meireles et al., 2016). Enzymes have many applications and are being used in industries like detergents, drugs, food & beverages, pulp, textiles, paper, and animal feeds (Vogel and May, 2019). These biological molecules may be classified into six classes; 1) oxidoreductases (like glucose oxidase, alcohol dehydrogenase, catalase, etc.) which catalyze the redox reaction and transfer oxygen or a hydrogen atom; 2) transferases (like lipid kinases, etc.) which allow the transfer of an atom or a group of atoms from one molecule to the next; 3) hydrolases (like glycosylase, aminopeptidase, etc.) which catalyze hydrolytic reactions; 4) lyases (like hydrogenase, pyruvate decarboxylase, synthase, etc.) which catalyze any reaction by removing an atom or group of atoms; 5) isomerases (like isomerase, epimerase, etc.) which catalyze the rearrangement in a molecule; and 6) ligases (e.g., synthetase, etc.) which catalyze the joining of two molecules with a covalent bond (Figure 5) (Shen and Chou, 2007).

Over the past few years, the use of enzymes as anti-biofilm agents has increased. (Taraszkiewicz et al., 2013; Thallinger et al., 2013). Many types of enzymes (Table 3) can be applied to eradicate biofilms from surfaces and thus reduce the problems associated with biofilms (Srey et al., 2013). Regulatory authorities have approved enzymes for the cleaning of food contact surfaces and there is no evidence that these enzymes interfere with the quality of food (Torres et al., 2019; Khadre et al., 2001).

Table 3. Anti-biofilm enzymes

Enzymes class	Enzymes applied	Biofilm source	Surface material	Effect	References
Hydrolase	Pandion, resinase, spenzyme, and paradigm used separately	P. aeruginosa	Polystyrene	Biofilm eradication	(Augustin et al., 2004)
Hydrolase	Bacteriophage enzyme	E. coli	Stainless steel	Biofilm removal	(Sharma et al., 2005)
Hydrolase	Lactonase	P. aeruginosa	Polystyrene	77% biofilm eradication	(Kiran et al., 2011)
Hydrolase	Bacteriophage enzyme	E. coli	Plastic pegs	99.99% biofilm eradication	(Lu and Collins, 2007)
Hydrolase	Pronase	Pseudomonas fluorescence	Borosilicate glass	Biofilm eradication	(Orgaz et al., 2007)
Hydrolase	Acylase	Bacterial sp.	Reverse osmosis membrane	9.0% biofilm eradication	(Kim et al., 2013)
Hydrolase	Lactonase by an engineered T7 bacteriophage	P. aeruginosa & E. coli	Polyvinyl chloride	Biofilm inhibition	(Pei and Lamas-Samanamud, 2014)
Hydrolase	Savinase®	Pseudomonas sp.	Polystyrene	Complete biofilm eradication	(Leroy et al., 2008)
Hydrolase	Savinase®	P. fluorescence	Glass wool	80% biofilm eradication	(Molobela et al., 2010)
Hydrolase	DNase	E. faecalis	Polystyrene	Biofilm eradication	(Thomas et al., 2008)
Hydrolase	Endolysin	S. aureus	Polystyrene	Biofilm removal	(Giaouris et al., 2015)
Hydrolase	DNase	L. monocytogenes	Polystyrene	50% biofilm eradication	(Nguyen and Burrows, 2014)
Hydrolase	Acylase 1 + proteinase K	Bacterial sp.	Reverse osmosis membrane	34% biofilm eradication	(Kim et al., 2013)
Hydrolase	Cellulase + Pronase	P. fluorescence	Borosilicate glass	94% biofilm removal	(Orgaz et al., 2007)

Table 3. (Continued)

Enzymes class	Enzymes applied	Biofilm source	Surface material	Effect	References
Hydrolase	DispersinB	Staphylococcus epidermidis	Glass	40% biofilm removal	(Brindle et al., 2011)
Hydrolase	Savinase® + sheer stress	P. aeruginosa	Polyethylene	90% biofilm eradication	(Pechaud et al., 2012)
Hydrolase	Amyloglucosidase	E. coli	Stainless steel	96% biofilm eradication	(Oulahal-Lagsir et al., 2003)
Hydrolase	Alpha-amylase	S. aureus, S. epidermidis	Polystyrene	79% biofilm eradication	(Craigen et al., 2011)
Lyase	Alginate lyase + gentamycin	P. aeruginosa	Cellulose fibers	Complete biofilm removal	(Alkawash et al., 2006)
Hydrolase	Dispersin B	S. epidermidis	Human cells	98%	(Pavlukhina et al., 2012)

Anti-QS Enzymes

The close association of microbial cells in biofilms help cells to coexist and the matrix provides optimum conditions for the Quorum sensing (QS) mechanism (Giaouris et al., 2015). QS is a complex regulatory system that relies on the production, release, and detection of small cell signaling molecules which are known as autoinducers (LaSarre and Federle, 2013). Many classes of signaling molecules have been reported but the three that have been extensively studied are autoinducers (AI), autoinducing peptides, and acyl-homoserine lactones (Miller and Bassler, 2001). The QS mechanisms have three main components: the AI, the gene that codes for AI synthase, and the gene that codes for response regulator proteins. Any anti-QS strategy requires the inhibition of one of these components (Kalia, 2013). Anti-QS enzymes like N-acyl homoserine lactonases and acylases can be potential candidates to inhibit biofilm formation (Lazar, 2011). Lactonases break down the bond of the homoserine ring, thus inhibiting the binding of homoserine lactones to transcriptional regulators (Thallinger et al., 2013). In a study, lactonases were able to reduce the biofilm of *P. aeruginosa* by up to 77% as well reducing the virulence factor (Kiran et al., 2011). Lactonase produced by the engineered T7 bacteriophage showed broad-spectrum anti-QS activity by inhibiting the biofilm of mixed-species of *P. aeruginosa* and *E. coli* (Figure 5) (Pei and Lamas-Samanamud, 2014).

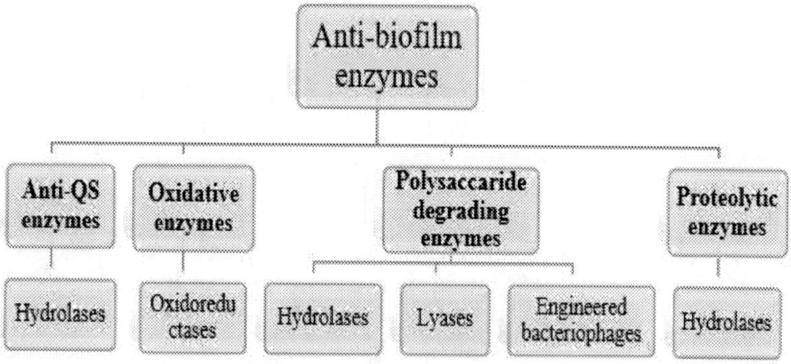

Figure 5. Important enzymes for the inhibition and control of biofilms.

Anti-Adhesion Strategies as Antibiofilm Approaches

Antiadhesion approaches can also play role in biofilm eradication or biofilm formation. This can lead to general or specific inhibition of adhesion depending on the target. By modifying the chemistry of the surface nonspecific inhibition of adhesion can be obtained (Beloin et al., 2014; Neoh et al., 2017). Micro or nanoscale engineering of surface topography is an advanced approach as it's non-toxic and does not depend on material type. One more advantage of this approach is that it can be modified chemically (Graham and Cady, 2014). There is not any concern regarding negative effects of nanoscale topographical modification (Liu et al., 2016; Cheng et al., 2019) however, more research is needed. The effect of surface modification has been studied on *C. albicans* biofilm formation. Poly dimethyl siloxane of different sizes was coated on surfaces and exposed to biofilm formation. Biofilm formation was found to be higher on the surfaces which were coated with 4-8μm while lower on the surfaces which were coated with 0.5-5μm particles (Lagree et al., 2018). In another study, biofilm formation of *B. subtilis*, *E. coli,* and *S. epidermidis* was inhibited up to 45% on surfaces coated with polydimethylsiloxane as compared to smooth surfaces (Perera-Costa et al., 2014).

Lower biofilm formation on topographically modified surfaces may be due to fewer binding sites on these surfaces as compared with smooth surfaces. The solid surface can also trap air which results in lower adhesion of microbes with that surface (Lagree et al., 2018). Various antiadhesive materials including synthetic polyethylene glycol and non-synthetic polymers like hyaluronic acid and agarose have been reported to have promising antiadhesive activities (Neoh et al., 2017; Zhou, 2014; Beloin et al., 2014; Junter et al., 2016). Targeting specific adhesion has shown strong antiadhesive potential. *E. coli* biofilm formation and adhesion were reduced to 90% in the presence of pilicides which target curli fimbriae and P pili. Similarly biofilm formation and adhesion which depends on curli was inhibited in the presence of curlicides (Cegelski et al., 2009). FimH inhibitor inhibits biofilm formation of multidrug-resistant *E. coli* which causes acute urinary tract infections (Totsika et al., 2013).

Conclusions

Biofilms are complex and diverse communities of microorganisms. These community structures protect microbes from harmful effects of the environment and increase their survival rate. Biofilms cause many types of problems in medicine; they increase the pathogenicity of microbes and cause implant failure. The biofilm increases the resistance of microbial cells to antibiotics. Scientists are looking for new approaches to control biofilms by adopting new techniques. Various approaches are emerging rapidly. Nanoparticles and NP based products can inhibit and eradicate biofilms. Targeted natural product delivery using nanocarriers can overcome biofilm-associated infections and problems. Further, because of the diverse mode of action of NPs, microbes cannot develop resistance against them. Nanoparticle-coated implants can be resistant to biofilm formation and thus reduce the rate of implant failure. Antimicrobial peptides are another efficient approach to handling biofilms. More than 5000 AMPs have been isolated to date from different sources. Almost all of these possess antimicrobial and antibiofilm potential. Like NPs, there are very rare chances of microbes developing resistance against AMPs. AMPs are natural peptides of the innate immune system and have been plying their antimicrobial role throughout evolution helping the immune system protect the body from microbial attack. Scientists have also developed synthetic AMPs which show good antimicrobial activity. Thus, AMPs can be used in advanced antibiofilm approaches. One potential approach is to inhibit microbial adhesion on the targeted surface. This will significantly reduce biofilm formation and ultimately reduce biofilm-associated problems. Inhibiting the QS mechanism can also reduce biofilm formation. The presence of anti-QS agents there will reduce or eliminate communication between microbes, and this will prevent biofilm formation. Antibiofilm enzymes can also perform their role in the inhibition of biofilms, especially in non-living systems. It can be concluded that there are a number of exciting new antibiofilm techniques. However, there is a need to evaluate these approaches in the appropriate settings.

References

Abhilash, M. (2010). Insilico analysis of cranberry *Proanthocyanidin epicatechin* (4beta-8, 2beta-0-7) as an inhibitor for modelled afimbrial adhesin virulence protein of uropathogenic *Escherichia coli. International Journal of Pharma and Bio Sciences*, 1-7.

Acharya, S., Hill, J. P. and Ariga, K. (2009). Soft Langmuir–Blodgett technique for hard nanomaterials. *Advanced Materials*, 2959-2981.

Adedeji, W. (2016). The treasure called antibiotics. *Annals of Ibadan postgraduate medicine*, 56.

Al-Qattan, M. N., Deb, P. K. and Tekade, R. K. (2018). Molecular dynamics simulation strategies for designing carbon-nanotube-based targeted drug delivery. *Drug Discovery Today*, 235-250.

Alanazi, F. K., Radwan, A. A. and Alsarra, I. A. (2010). Biopharmaceutical applications of nanogold. *Saudi Pharmaceutical Journal*, 179-193.

Alaqad, K. and Saleh, T. A. (2016). Gold and silver nanoparticles: synthesis methods, characterization routes and applications towards drugs. *Journal of Environmental & Analytical Toxicology*, 525-2161.

Ali, K., Ahmed, B., Ansari, S. M., Saquib, Q., Al-Khedhairy, A. A., Dwivedi, S., Alshaeri, M., Khan, M. S. and Musarrat, J. (2019). Comparative in situ ROS mediated killing of bacteria with bulk analogue, Eucalyptus leaf extract (ELE)-capped and bare surface copper oxide nanoparticles. *Materials Science and Engineering: C*, 747-758.

Alkawash, M. A., Soothill, J. S. and Schiller, N. L. (2006). Alginate lyase enhances antibiotic killing of mucoid *Pseudomonas aeruginosa* in biofilms. *Apmis*, 131-138.

Andersson, D. I., Hughes, D. and Kubicek-Sutherland, J. Z. (2016). Mechanisms and consequences of bacterial resistance to antimicrobial peptides. *Drug Resistance Updates*, 43-57.

Anghel, A. G., Grumezescu, A. M., Chirea, M., Grumezescu, V., Socol, G., Iordache, F., Oprea, A. E., Anghel, I. and Holban, A. M. (2014). MAPLE fabricated Fe_3O_4@ *Cinnamomum verum* antimicrobial surfaces for improved gastrostomy tubes. *Molecules*, 8981-8994.

Augustin, M., Ali-Vehmas, T. and Atroshi, F. (2004). Assessment of enzymatic cleaning agents and disinfectants against bacterial biofilms.

Avramović, N., Mandić, B., Savić-Radojević, A. and Simić, T. (2020). Polymeric nanocarriers of drug delivery systems in cancer therapy. *Pharmaceutics*, 298.

Bahar, A. A. and Ren, D. (2013). Antimicrobial peptides. *Pharmaceuticals*, 1543-1575.

Barros, C. H., Devlin, H., Hiebner, D. W., Vitale, S., Quinn, L. and Casey, E. (2020). Enhancing curcumin's solubility and antibiofilm activity via silica surface modification. *Nanoscale Advances*, 1694-1708.

Battaglia, L. and Gallarate, M. (2012). Lipid nanoparticles: state of the art, new preparation methods and challenges in drug delivery. *Expert opinion on drug delivery*, 497-508.

Beloin, C., Renard, S., Ghigo, J.-M. and Lebeaux, D. (2014). Novel approaches to combat bacterial biofilms. *Current opinion in pharmacology*, 61-68.

Blecher, K., Nasir, A. and Friedman, A. (2011). The growing role of nanotechnology in combating infectious disease. *Virulence*, 395-401.

Brindle, E. R., Miller, D. A. and Stewart, P. S. (2011). Hydrodynamic deformation and removal of *Staphylococcus epidermidis* biofilms treated with urea, chlorhexidine, iron chloride, or DispersinB. *Biotechnology and bioengineering*, 2968-2977.

Brogden, K. A. (2005). Antimicrobial peptides: pore formers or metabolic inhibitors in bacteria? *Nature reviews microbiology*, 238-250.

Campoccia, D., Montanaro, L. and Arciola, C. R. (2013). A review of the clinical implications of anti-infective biomaterials and infection-resistant surfaces. *Biomaterials*, 8018-8029.

Castillo-Juárez, I., García-Contreras, R., Velázquez-Guadarrama, N., Soto-Hernández, M. and Martínez-Vázquez, M. (2013). Amphypterygium adstringens anacardic acid mixture inhibits quorum sensing-controlled virulence factors of *Chromobacterium violaceum* and *Pseudomonas aeruginosa*. *Archives of medical research*, 488-494.

Cegelski, L., Pinkner, J. S., Hammer, N. D., Cusumano, C. K., Hung, C. S., Chorell, E., Åberg, V., Walker, J. N., Seed, P. C. and Almqvist, F. (2009). Small-molecule inhibitors target *Escherichia coli* amyloid biogenesis and biofilm formation. *Nature chemical biology*, 913-919.

Chamundeeswari, M., Sobhana, S. L., Jacob, J. P., Kumar, M. G., Devi, M. P., Sastry, T. P. and Mandal, A. B. (2010). Preparation, characterization and evaluation of a biopolymeric gold nanocomposite with antimicrobial activity. *Biotechnology and applied biochemistry*, 29-35.

Chen, Y., Feng, C., Zhang, Q., Luo, M., Xu, J. and Han, Q. (2020). Engineering of antibacterial/recyclable difunctional nanoparticles via synergism of quaternary ammonia salt site and N-halamine sites on magnetic surface. *Colloids and surfaces B: Biointerfaces*, 110642.

Cheng, Y., Feng, G. and Moraru, C. I. (2019). Micro-and nanotopography sensitive bacterial attachment mechanisms: a review. *Frontiers in microbiology*, 191.

Comin, V. M., Lopes, L. Q., Quatrin, P. M., de Souza, M. E., Bonez, P. C., Pintos, F. G., Raffin, R. P., Vaucher, R. d. A., Martinez, D. S. and Santos, R. C. (2016). Influence of *Melaleuca alternifolia* oil nanoparticles on aspects of *Pseudomonas aeruginosa* biofilm. *Microbial pathogenesis*, 120-125.

Corrêa, J. A. F., Evangelista, A. G., de Melo Nazareth, T. and Luciano, F. B. (2019). Fundamentals on the molecular mechanism of action of antimicrobial peptides. *Materialia*, 100494.

Craigen, B., Dashiff, A. and Kadouri, D. E. (2011). The use of commercially available alpha-amylase compounds to inhibit and remove *Staphylococcus aureus* biofilms. *The open microbiology journal*, 21.

Cui, L., Chen, P., Chen, S., Yuan, Z., Yu, C., Ren, B. and Zhang, K. (2013). In situ study of the antibacterial activity and mechanism of action of silver nanoparticles by surface-enhanced Raman spectroscopy. *Analytical chemistry*, 5436-5443.

da Silva Gündel, S., de Souza, M. E., Quatrin, P. M., Klein, B., Wagner, R., Gündel, A., de Almeida Vaucher, R., Santos, R. C. V. and Ourique, A. F. (2018). Nanoemulsions containing Cymbopogon flexuosus essential oil: Development, characterization, stability study and evaluation of antimicrobial and antibiofilm activities. *Microbial pathogenesis*, 268-276.

Dakal, T. C., Kumar, A., Majumdar, R. S. and Yadav, V. (2016). Mechanistic basis of antimicrobial actions of silver nanoparticles. *Frontiers in microbiology*, 1831.

Dastjerdi, R. and Montazer, M. (2010). A review on the application of inorganic nano-structured materials in the modification of textiles: focus on anti-microbial properties. *Colloids and surfaces B: Biointerfaces*, 5-18.

de Barros, C. H. N., Cruz, G. C. F., Mayrink, W. and Tasic, L. (2018). Bio-based synthesis of silver nanoparticles from orange waste: effects of distinct biomolecule coatings on size, morphology, and antimicrobial activity. *Nanotechnology, science and applications*, 1.

de Kraker, M. E., Stewardson, A. J. and Harbarth, S. (2016). Will 10 million people die a year due to antimicrobial resistance by 2050? *PLoS medicine*, e1002184.

De La Fuente-Núñez, C., Korolik, V., Bains, M., Nguyen, U., Breidenstein, E. B., Horsman, S., Lewenza, S., Burrows, L. and Hancock, R. E. (2012). Inhibition of bacterial biofilm formation and swarming motility by a small synthetic cationic peptide. *Antimicrobial agents and chemotherapy*, 2696-2704.

de Mélo Silva, I. S., Gaspar, L. M. d. A. C., Rocha, A. M. O., da Costa, L. P., Tada, D. B., Franceschi, E. and Padilha, F. F. (2020). Encapsulation of red propolis in polymer nanoparticles for the destruction of pathogenic biofilms. *Aaps Pharmscitech*, 1-9.

Dean, S. N., Bishop, B. M. and Van Hoek, M. L. (2011). Natural and synthetic cathelicidin peptides with anti-microbial and anti-biofilm activity against *Staphylococcus aureus*. *BMC microbiology*, 1-13.

del Carpio-Perochena, A., Kishen, A., Felitti, R., Bhagirath, A. Y., Medapati, M. R., Lai, C. and Cunha, R. S. (2017). Antibacterial properties of chitosan nanoparticles and propolis associated with calcium hydroxide against single-and multispecies biofilms: an *in vitro* and *in situ* study. *Journal of Endodontics*, 1332-1336.

Donlan, R. M. (2001). Biofilm formation: a clinically relevant microbiological process. *Clinical infectious diseases*, 1387-1392.

El-Guendouz, S., Aazza, S., Lyoussi, B., Bankova, V., Lourenço, J. P., Costa, A. M. R., Mariano, J. F., Miguel, M. G. and Faleiro, M. L. (2016). Impact of biohybrid magnetite nanoparticles and moroccan propolis on adherence of methicillin resistant strains of *Staphylococcus aureus*. *Molecules*, 1208.

Elbialy, N. S., Aboushoushah, S. F., Sofi, B. F. and Noorwali, A. (2020). Multifunctional curcumin-loaded mesoporous silica nanoparticles for cancer chemoprevention and therapy. *Microporous and Mesoporous Materials*, 109540.

Epand, R. M., Walker, C., Epand, R. F. and Magarvey, N. A. (2016). Molecular mechanisms of membrane targeting antibiotics. *Biochimica et Biophysica Acta (BBA)-Biomembranes*, 980-987.

Ge, X., Wei, M., He, S. and Yuan, W. E. (2019). Advances of non-ionic surfactant vesicles (niosomes) and their application in drug delivery. *Pharmaceutics*, 55.

Ghorbanzadeh, R., Assadian, H., Chiniforush, N., Parker, S., Pourakbari, B., Ehsani, B., Alikhani, M. Y. and Bahador, A. (2020). Modulation of virulence in *Enterococcus faecalis* cells surviving antimicrobial photodynamic inactivation with reduced graphene oxide-curcumin: An *ex vivo* biofilm model. *Photodiagnosis and photodynamic therapy*, 101643.

Giaouris, E., Heir, E., Desvaux, M., Hébraud, M., Møretrø, T., Langsrud, S., Doulgeraki, A., Nychas, G. J., Kačániová, M. and Czaczyk, K. (2015). Intra-and inter-species interactions within biofilms of important foodborne bacterial pathogens. *Frontiers in microbiology*, 841.

Girish, V. M., Liang, H., Aguilan, J. T., Nosanchuk, J. D., Friedman, J. M. and Nacharaju, P. (2019). Anti-biofilm activity of garlic extract loaded nanoparticles. *Nanomedicine: Nanotechnology, Biology and Medicine*, 102009.

Graham, M. V. and Cady, N. C. (2014). Nano and microscale topographies for the prevention of bacterial surface fouling. *Coatings*, 37-59.

Grenho, L., Barros, J., Ferreira, C., Santos, V., Monteiro, F., Ferraz, M. P. and Cortes, M. (2015). *In vitro* antimicrobial activity and biocompatibility of propolis containing nanohydroxyapatite. *Biomedical materials*, 025004.

Grumezescu, V., Holban, A. M., Iordache, F., Socol, G., Mogoşanu, G. D., Grumezescu, A. M., Ficai, A., Vasile, B. Ş., Truşcă, R. and Chifiriuc, M. C. (2014). MAPLE fabricated magnetite@ eugenol and (3-hidroxybutyric acid-co-3-hidroxyvaleric acid)–polyvinyl alcohol microspheres coated surfaces with anti-microbial properties. *Applied surface science*, 16-22.

Guilhelmelli, F., Vilela, N., Albuquerque, P., Derengowski, L., Silva-Pereira, I. and Kyaw, C. (2013). Antibiotic development challenges: the various mechanisms of action of antimicrobial peptides and of bacterial resistance. *Frontiers in microbiology*, 353.

Haney, E. F., Straus, S. K. and Hancock, R. E. (2019). Reassessing the host defense peptide landscape. *Frontiers in Chemistry*, 43.

Haruna, K., Saleh, T. A., Al Thagfi, J. and Al-Saadi, A. A. (2016). Structural properties, vibrational spectra and surface-enhanced Raman scattering of 2, 4, 6-trichloro-and tribromoanilines: a comparative study. *Journal of Molecular Structure*, 7-15.

Hasan, S. (2015). A review on nanoparticles: their synthesis and types. *Res. J. Recent Sci*, 2502.

Hendessi, S., Sevinis, E. B., Unal, S., Cebeci, F. C., Menceloglu, Y. Z. and Unal, H. (2016). Antibacterial sustained-release coatings from halloysite nanotubes/ waterborne polyurethanes. *Progress in Organic Coatings*, 253-261.

Hilchie, A. L., Wuerth, K. and Hancock, R. E. (2013). Immune modulation by multifaceted cationic host defense (antimicrobial) peptides. *Nature Chemical Biology*, 761-768.

Hindi, K. M., Ditto, A. J., Panzner, M. J., Medvetz, D. A., Han, D. S., Hovis, C. E., Hilliard, J. K., Taylor, J. B., Yun, Y. H. and Cannon, C. L. (2009). The antimicrobial efficacy of sustained release silver–carbene complex-loaded L-tyrosine polyphosphate nanoparticles: Characterization, *in vitro* and *in vivo* studies. *Biomaterials*, 3771-3779.

Hoseinnejad, M., Jafari, S. M. and Katouzian, I. (2018). Inorganic and metal nanoparticles and their antimicrobial activity in food packaging applications. *Critical reviews in microbiology*, 161-181.

Huang, T. C., Lee, J. F. and Chen, J. Y. (2011). Pardaxin, an antimicrobial peptide, triggers caspase-dependent and ROS-mediated apoptosis in HT-1080 cells. *Marine drugs*, 1995-2009.

Huang, Z., Zheng, X., Yan, D., Yin, G., Liao, X., Kang, Y., Yao, Y., Huang, D. and Hao, B. (2008). Toxicological effect of ZnO nanoparticles based on bacteria. *Langmuir*, 4140-4144.

Jahanizadeh, S., Yazdian, F., Marjani, A., Omidi, M. and Rashedi, H. (2017). Curcumin-loaded chitosan/carboxymethyl starch/montmorillonite bio-nanocomposite for reduction of dental bacterial biofilm formation. *International journal of biological macromolecules*, 757-763.

Jaiswal, S. and Mishra, P. (2018). Antimicrobial and antibiofilm activity of curcumin-silver nanoparticles with improved stability and selective toxicity to bacteria over mammalian cells. *Medical microbiology and immunology*, 39-53.

Jenssen, H., Hamill, P. and Hancock, R. E. (2006). Peptide antimicrobial agents. *Clinical microbiology reviews*, 491-511.

Johnston, H. and Hutchison, G. (2010). Christensen FM, Peters S., Hankin S., Stone V. *crit. rev. toxicol*, 328-346.

Junter, G. A., Thébault, P. and Lebrun, L. (2016). Polysaccharide-based antibiofilm surfaces. *Acta biomaterialia*, 13-25.

Kalia, V. C. (2013). Quorum sensing inhibitors: an overview. *Biotechnology advances*, 224-245.

Kang, S., Pinault, M., Pfefferle, L. D. and Elimelech, M. (2007). Single-walled carbon nanotubes exhibit strong antimicrobial activity. *Langmuir*, 8670-8673.

Kaur, G. and Mehta, S. (2017). Developments of Polysorbate (Tween) based microemulsions: Preclinical drug delivery, toxicity and antimicrobial applications. *International journal of pharmaceutics*, 134-160.

Kaweeteerawat, C., Na Ubol, P., Sangmuang, S., Aueviriyavit, S. and Maniratanachote, R. (2017). Mechanisms of antibiotic resistance in bacteria mediated by silver nanoparticles. *Journal of Toxicology and Environmental Health, Part A*, 1276-1289.

Khadre, M., Yousef, A. and Kim, J. G. (2001). Microbiological aspects of ozone applications in food: a review. *Journal of food science*, 1242-1252.

Kim, L. H., Kim, S. J., Kim, C. M., Shin, M. S., Kook, S. and Kim, I. S. (2013). Effects of enzymatic treatment on the reduction of extracellular polymeric substances (EPS) from biofouled membranes. *Desalination and Water Treatment*, 6355-6361.

Kiran, S., Sharma, P., Harjai, K. and Capalash, N. (2011). Enzymatic quorum quenching increases antibiotic susceptibility of multidrug resistant *Pseudomonas aeruginosa. Iranian Journal of Microbiology*, 1.

Koo, H. B. and Seo, J. (2019). Antimicrobial peptides under clinical investigation. *Peptide Science*, e24122.

Kumar, P., Kizhakkedathu, J. N. and Straus, S. K. (2018). Antimicrobial peptides: diversity, mechanism of action and strategies to improve the activity and biocompatibility in vivo. *Biomolecules*, 4.

Lagree, K., Mon, H. H., Mitchell, A. P. and Ducker, W. A. (2018). Impact of surface topography on biofilm formation by *Candida albicans. PLoS One*, e0197925.

Lara, H. H., Ayala-Núnez, N. V., Turrent, L. d. C. I. and Padilla, C. R. (2010). Bactericidal effect of silver nanoparticles against multidrug-resistant bacteria. *World Journal of Microbiology and Biotechnology*, 615-621.

LaRock, C. N. and Nizet, V. (2015). Cationic antimicrobial peptide resistance mechanisms of streptococcal pathogens. *Biochimica et Biophysica Acta (BBA)-Biomembranes*, 3047-3054.

LaSarre, B. and Federle, M. J. (2013). Exploiting quorum sensing to confuse bacterial pathogens. *Microbiology and molecular biology reviews*, 73-111.

Lazar, V. (2011). Quorum sensing in biofilms–how to destroy the bacterial citadels or their cohesion/power? *Anaerobe*, 280-285.

Le, C. F., Fang, C. M. and Sekaran, S. D. (2017). Intracellular targeting mechanisms by antimicrobial peptides. *Antimicrobial agents and chemotherapy*, e02340-02316.

Lee, T. H., N Hall, K. and Aguilar, M. I. (2016). Antimicrobial peptide structure and mechanism of action: a focus on the role of membrane structure. *Current topics in medicinal chemistry*, 25-39.

Leroy, C., Delbarre, C., Ghillebaert, F., Compere, C. and Combes, D. (2008). Effects of commercial enzymes on the adhesion of a marine biofilm-forming bacterium. *Biofouling*, 11-22.

Li, W. R., Xie, X. B., Shi, Q. S., Zeng, H. Y., You-Sheng, O. Y. and Chen, Y. B. (2010). Antibacterial activity and mechanism of silver nanoparticles on *Escherichia coli*. *Applied microbiology and biotechnology*, 1115-1122.

Liu, L., Ercan, B., Sun, L., Ziemer, K. S. and Webster, T. J. (2016). Understanding the role of polymer surface nanoscale topography on inhibiting bacteria adhesion and growth. *ACS Biomaterials Science & Engineering*, 122-130.

Lou, Z., Chen, J., Yu, F., Wang, H., Kou, X., Ma, C. and Zhu, S. (2017). The antioxidant, antibacterial, antibiofilm activity of essential oil from *Citrus medica* L. var. sarcodactylis and its nanoemulsion. *LWT*, 371-377.

Lu, T. K. and Collins, J. J. (2007). Dispersing biofilms with engineered enzymatic bacteriophage. *Proceedings of the National Academy of Sciences*, 11197-11202.

Mader, J. S. and Hoskin, D. W. (2006). Cationic antimicrobial peptides as novel cytotoxic agents for cancer treatment. *Expert opinion on investigational drugs*, 933-946.

Maghsoudi, A., Yazdian, F., Shahmoradi, S., Ghaderi, L., Hemati, M. and Amoabediny, G. (2017). Curcumin-loaded polysaccharide nanoparticles: Optimization and anticariogenic activity against *Streptococcus mutans*. *Materials Science and Engineering: C*, 1259-1267.

Malanovic, N. and Lohner, K. (2016). Antimicrobial peptides targeting Gram-positive bacteria. *Pharmaceuticals*, 59.

Maness, P. C., Smolinski, S., Blake, D. M., Huang, Z., Wolfrum, E. J. and Jacoby, W. A. (1999). Bactericidal activity of photocatalytic TiO2 reaction: toward an understanding of its killing mechanism. *Applied and environmental microbiology*, 4094-4098.

Manju, S., Malaikozhundan, B., Vijayakumar, S., Shanthi, S., Jaishabanu, A., Ekambaram, P. and Vaseeharan, B. (2016). Antibacterial, antibiofilm and cytotoxic effects of Nigella sativa essential oil coated gold nanoparticles. *Microbial pathogenesis*, 129-135.

Meireles, A., Borges, A., Giaouris, E. and Simões, M. (2016). The current knowledge on the application of anti-biofilm enzymes in the food industry. *Food Research International*, 140-146.

Menon, S., Rajeshkumar, S. and Kumar, V. (2017). A review on biogenic synthesis of gold nanoparticles, characterization, and its applications. *Resource-Efficient Technologies*, 516-527.

Michael, C. A., Dominey-Howes, D. and Labbate, M. (2014). The antimicrobial resistance crisis: causes, consequences, and management. *Frontiers in public health*, 145.

Miller, M. B. and Bassler, B. L. (2001). Quorum sensing in bacteria. *Annual Reviews in Microbiology*, 165-199.

Mirzahosseinipour, M., Khorsandi, K., Hosseinzadeh, R., Ghazaeian, M. and Shahidi, F. K. (2020). Antimicrobial photodynamic and wound healing activity of curcumin encapsulated in silica nanoparticles. *Photodiagnosis and photodynamic therapy*, 101639.

Moghimi, R., Aliahmadi, A., Rafati, H., Abtahi, H. R., Amini, S. and Feizabadi, M. M. (2018). Antibacterial and anti-biofilm activity of nanoemulsion of *Thymus daenensis* oil against multi-drug resistant *Acinetobacter baumannii. Journal of Molecular Liquids*, 765-770.

Molobela, I. P., Cloete, T. E. and Beukes, M. (2010). Protease and amylase enzymes for biofilm removal and degradation of extracellular polymeric substances (EPS) produced by Pseudomonas fluorescens bacteria.

Mookherjee, N., Anderson, M. A., Haagsman, H. P. and Davidson, D. J. (2020). Antimicrobial host defence peptides: functions and clinical potential. *Nature Reviews Drug Discovery*, 311-332.

Müller, R. H., Mäder, K. and Gohla, S. (2000). Solid lipid nanoparticles (SLN) for controlled drug delivery–a review of the state of the art. *European journal of pharmaceutics and biopharmaceutics*, 161-177.

Münch, D. and Sahl, H. G. (2015). Structural variations of the cell wall precursor lipid II in Gram-positive bacteria—Impact on binding and efficacy of antimicrobial peptides. *Biochimica et Biophysica Acta (BBA)-Biomembranes*, 3062-3071.

Mwangi, J., Hao, X., Lai, R. and Zhang, Z. Y. (2019). Antimicrobial peptides: new hope in the war against multidrug resistance. *Zoological research*, 488.

Neoh, K. G., Li, M., Kang, E. T., Chiong, E. and Tambyah, P. A. (2017). Surface modification strategies for combating catheter-related complications: recent advances and challenges. *Journal of Materials Chemistry B*, 2045-2067.

Nguyen, U. T. and Burrows, L. L. (2014). DNase I and proteinase K impair *Listeria monocytogenes* biofilm formation and induce dispersal of pre-existing biofilms. *International journal of food microbiology*, 26-32.

Niyonsaba, F., Iwabuchi, K., Someya, A., Hirata, M., Matsuda, H., Ogawa, H. and Nagaoka, I. (2002). A cathelicidin family of human antibacterial peptide LL-37 induces mast cell chemotaxis. *Immunology*, 20-26.

Ong, T. H., Chitra, E., Ramamurthy, S., Ling, C. C. S., Ambu, S. P. and Davamani, F. (2019). Cationic chitosan-propolis nanoparticles alter the zeta potential of *S. epidermidis*, inhibit biofilm formation by modulating gene expression and exhibit synergism with antibiotics. *PLoS One*, e0213079.

Ong, T. H., Chitra, E., Ramamurthy, S., Siddalingam, R. P., Yuen, K. H., Ambu, S. P. and Davamani, F. (2017). Chitosan-propolis nanoparticle formulation demonstrates anti-bacterial activity against Enterococcus faecalis biofilms. *PLoS One*, e0174888.

Orgaz, B., Neufeld, R. and SanJose, C. (2007). Single-step biofilm removal with delayed release encapsulated Pronase mixed with soluble enzymes. *Enzyme and Microbial Technology*, 1045-1051.

Otto, M. (2013). Staphylococcal infections: mechanisms of biofilm maturation and detachment as critical determinants of pathogenicity. *Annual review of medicine*, 175-188.

Oulahal Lagsir, N., Martial-Gros, A., Bonneau, M. and Blum, L. J. (2003). "*Escherichia coli* milk" biofilm removal from stainless steel surfaces: synergism between ultrasonic waves and enzymes. *Biofueling*, 159-168.

Overhage, J., Campisano, A., Bains, M., Torfs, E. C., Rehm, B. H. and Hancock, R. E. (2008). Human host defense peptide LL-37 prevents bacterial biofilm formation. *Infection and immunity*, 4176-4182.

Palmieri, V., Bugli, F., Cacaci, M., Perini, G., Maio, F. D., Delogu, G., Torelli, R., Conti, C., Sanguinetti, M. and Spirito, M. D. (2018). Graphene oxide coatings prevent *Candida albicans* biofilm formation with a controlled release of curcumin-loaded nanocomposites. *Nanomedicine*, 2867-2879.

Park, Y. K. and Hahm, K. S. (2005). Antimicrobial peptides (AMPs): peptide structure and mode of action. *BMB Reports*, 507-516.

Patrzykat, A., Friedrich, C. L., Zhang, L., Mendoza, V. and Hancock, R. E. (2002). Sublethal concentrations of pleurocidin-derived antimicrobial peptides inhibit macromolecular synthesis in *Escherichia coli*. *Antimicrobial agents and chemotherapy*, 605-614.

Pavlukhina, S. V., Kaplan, J. B., Xu, L., Chang, W., Yu, X., Madhyastha, S., Yakandawala, N., Mentbayeva, A., Khan, B. and Sukhishvili, S. A. (2012). Noneluting enzymatic antibiofilm coatings. *ACS applied materials & interfaces*, 4708-4716.

Pavoni, L., Perinelli, D. R., Bonacucina, G., Cespi, M. and Palmieri, G. F. (2020). An overview of micro-and nanoemulsions as vehicles for essential oils: Formulation, preparation and stability. *Nanomaterials*, 135.

Pechaud, Y., Marcato-Romain, C. E., Girbal-Neuhauser, E., Queinnec, I., Bessiere, Y. and Paul, E. (2012). Combining hydrodynamic and enzymatic treatments to improve multi-species thick biofilm removal. *Chemical engineering science*, 109-118.

Pei, R. and Lamas-Samanamud, G. R. (2014). Inhibition of biofilm formation by T7 bacteriophages producing quorum-quenching enzymes. *Applied and environmental microbiology*, 5340-5348.

Peng, K. T., Chiang, Y. C., Huang, T. Y., Chen, P. C., Chang, P. J. and Lee, C. W. (2019). Curcumin nanoparticles are a promising anti-bacterial and anti-inflammatory agent for treating periprosthetic joint infections. *International journal of nanomedicine*, 469.

Perera-Costa, D., Bruque, J. M., González-Martín, M. a. L., Gómez-García, A. C. and Vadillo-Rodriguez, V. (2014). Studying the influence of surface topography on bacterial adhesion using spatially organized microtopographic surface patterns. *Langmuir*, 4633-4641.

Perez, A. P., Perez, N., Lozano, C. M. S., Altube, M. J., de Farias, M. A., Portugal, R. V., Buzzola, F., Morilla, M. J. and Romero, E. L. (2019). The anti MRSA biofilm activity of *Thymus vulgaris* essential oil in nanovesicles. *Phytomedicine*, 339-351.

Pourhajibagher, M., Parker, S., Chiniforush, N. and Bahador, A. (2019). Photoexcitation triggering via semiconductor Graphene Quantum Dots by photochemical doping with Curcumin versus perio-pathogens mixed biofilms. *Photodiagnosis and photodynamic therapy*, 125-131.

Pourhajibagher, M., Plotino, G., Chiniforush, N. and Bahador, A. (2020). Dual wavelength irradiation antimicrobial photodynamic therapy using indocyanine green and metformin doped with nano-curcumin as an efficient adjunctive endodontic treatment modality. *Photodiagnosis and photodynamic therapy*, 101628.

Prakash, A., Vadivel, V., Rubini, D. and Nithyanand, P. (2019). Antibacterial and antibiofilm activities of linalool nanoemulsions against *Salmonella typhimurium*. *Food bioscience*, 57-65.

Qi, L., Xu, Z., Jiang, X., Hu, C. and Zou, X. (2004). Preparation and antibacterial activity of chitosan nanoparticles. *Carbohydrate research*, 2693-2700.

Rahman, M. R. T., Lou, Z., Yu, F., Wang, P. and Wang, H. (2017). Anti-quorum sensing and anti-biofilm activity of Amomum tsaoko (Amommum tsao-ko Crevost et Lemarie) on foodborne pathogens. *Saudi journal of biological sciences*, 324-330.

Ramasamy, M., Lee, J. H. and Lee, J. (2017a). Development of gold nanoparticles coated with silica containing the antibiofilm drug cinnamaldehyde and their effects on pathogenic bacteria. *International journal of nanomedicine*, 2813.

Ramasamy, M., Lee, J. H. and Lee, J. (2017b). Direct one-pot synthesis of cinnamaldehyde immobilized on gold nanoparticles and their antibiofilm properties. *Colloids and surfaces B: Biointerfaces*, 639-648.

Raza, A., Sime, F. B., Cabot, P. J., Maqbool, F., Roberts, J. A. and Falconer, J. R. (2019). Solid nanoparticles for oral antimicrobial drug delivery: A review. *Drug Discovery Today*, 858-866.

Ren, G., Hu, D., Cheng, E. W., Vargas-Reus, M. A., Reip, P. and Allaker, R. P. (2009). Characterisation of copper oxide nanoparticles for antimicrobial applications. *International journal of antimicrobial agents*, 587-590.

Ribeiro, K. L., Frías, I. A., Franco, O. L., Dias, S. C., Sousa-Junior, A. A., Silva, O. N., Bakuzis, A. F., Oliveira, M. D. and Andrade, C. A. (2018a). Clavanin A-bioconjugated Fe_3O_4/Silane core-shell nanoparticles for thermal ablation of bacterial biofilms. *Colloids and surfaces B: Biointerfaces*, 72-81.

Ribeiro, K. L., Frías, I. A., Franco, O. L., Dias, S. C., Sousa-Junior, A. A., Silva, O. N., Bakuzis, A. F., Oliveira, M. D. and Andrade, C. A. (2018b). Clavanin A-bioconjugated Fe_3O_4/Silane core-shell nanoparticles for thermal ablation of bacterial biofilms B Biointerfaces.

Rigby, S., Fairhead, M. and van der Walle, C. F. (2008). Engineering silica particles as oral drug delivery vehicles. *Current pharmaceutical design*, 1821-1831.

Rossi, F., Diaz, L., Wollam, A., Panesso, D., Zhou, Y., Rincon, S., Narechania, A., Xing, G., Di Gioia, T. S. and Doi, A. (2014). Transferable vancomycin resistance in a community-associated MRSA lineage. *New England Journal of Medicine*, 1524-1531.

Rossi, G. G., Guterres, K. B., Bonez, P. C., da Silva Gundel, S., Aggertt, V. A., Siqueira, F. S., Ourique, A. F., Wagnerd, R., Klein, B. and Santos, R. C. V. (2017). Antibiofilm activity of nanoemulsions of *Cymbopogon flexuosus* against rapidly growing mycobacteria. *Microbial pathogenesis*, 335-341.

Sadredinamin, M., Mehrnejad, F., Hosseini, P. and Doustdar, F. (2016). Antimicrobial Peptides (AMPs). *Novelty in Biomedicine*, 70-76.

Shahverdi, A. R., Fakhimi, A., Shahverdi, H. R. and Minaian, S. (2007). Synthesis and effect of silver nanoparticles on the antibacterial activity of different antibiotics against *Staphylococcus aureus* and *Escherichia coli*. *Nanomedicine: Nanotechnology, Biology and Medicine*, 168-171.

Shai, Y. (2002). Mode of action of membrane active antimicrobial peptides. *Peptide Science: Original Research on Biomolecules*, 236-248.

Shai, Y. and Oren, Z. (2001). From "carpet" mechanism to de-novo designed diastereomeric cell-selective antimicrobial peptides. *Peptides*, 1629-1641.

Shanmugam, S. (2009). *Enzyme technology*. IK International Pvt Ltd.

Shariati, A., Asadian, E., Fallah, F., Azimi, T., Hashemi, A., Sharahi, J. Y. and Moghadam, M. T. (2019). Evaluation of Nano-curcumin effects on expression levels of virulence genes and biofilm production of multidrug-resistant *Pseudomonas aeruginosa* isolated from burn wound infection in Tehran, Iran. *Infection and drug resistance*, 2223.

Sharma, D., Kanchi, S. and Bisetty, K. (2019). Biogenic synthesis of nanoparticles: a review. *Arabian journal of chemistry*, 3576-3600.

Sharma, M., Ryu, J. H. and Beuchat, L. (2005). Inactivation of *Escherichia coli* O157: H7 in biofilm on stainless steel by treatment with an alkaline cleaner and a bacteriophage. *Journal of applied microbiology*, 449-459.

Shen, H. B. and Chou, K. C. (2007). EzyPred: a top–down approach for predicting enzyme functional classes and subclasses. *Biochemical and biophysical research communications*, 53-59.

Shirodkar, R. K., Kumar, L., Mutalik, S. and Lewis, S. (2019). Solid lipid nanoparticles and nanostructured lipid carriers: emerging lipid based drug delivery systems. *Pharmaceutical Chemistry Journal*, 440-453.

Simons, A., Alhanout, K. and Duval, R. E. (2020). Bacteriocins, antimicrobial peptides from bacterial origin: overview of their biology and their impact against multidrug-resistant bacteria. *Microorganisms*, 639.

Singh, A. K., Prakash, P., Singh, R., Nandy, N., Firdaus, Z., Bansal, M., Singh, R. K., Srivastava, A., Roy, J. K. and Mishra, B. (2017). Curcumin quantum dots mediated degradation of bacterial biofilms. *Frontiers in microbiology*, 1517.

Singh, B. N., Pandey, G., Jadaun, V., Singh, S., Bajpai, R., Nayaka, S., Naqvi, A. H., Rawat, A. K. S., Upreti, D. K. and Singh, B. R. (2015). Development and characterization of a novel Swarna-based herbo-metallic colloidal nano-formulation–inhibitor of *Streptococcus mutans* quorum sensing. *RSC Advances*, 5809-5822.

Singh, S., Singh, S. K., Chowdhury, I. and Singh, R. (2017). Understanding the mechanism of bacterial biofilms resistance to antimicrobial agents. *The open microbiology journal*, 53.

Singh, V. K., Mishra, A. and Jha, B. (2017). Anti-quorum sensing and anti-biofilm activity of Delftia tsuruhatensis extract by attenuating the quorum sensing-controlled virulence factor production in *Pseudomonas aeruginosa*. *Frontiers in cellular and infection microbiology*, 337.

Souza, M., Lopes, L., Bonez, P., Gündel, A., Martinez, D., Sagrillo, M., Giongo, J., Vaucher, R., Raffin, R. and Boligon, A. (2017). *Melaleuca alternifolia* nanoparticles against *Candida* species biofilms. *Microbial pathogenesis*, 125-132.

Srey, S., Jahid, I. K. and Ha, S.-D. (2013). Biofilm formation in food industries: a food safety concern. *Food control*, 572-585.

Stan, M. S., Constanda, S., Grumezescu, V., Andronescu, E., Ene, A. M., Holban, A. M., Vasile, B. S., Mogoantă, L., Bălşeanu, T.-A. and Mogoşanu, G. D. (2016). Thin coatings based on ZnO- C18-usnic acid nanoparticles prepared by MAPLE inhibit the development of *Salmonella enterica* early biofilm growth. *Applied surface science*, 318-325.

Tang, F., Li, L. and Chen, D. (2012). Mesoporous silica nanoparticles: synthesis, biocompatibility and drug delivery. *Advanced Materials*, 1504-1534.

Taraszkiewicz, A., Fila, G., Grinholc, M. and Nakonieczna, J. (2013). Innovative strategies to overcome biofilm resistance. *BioMed Research International*.

Thallinger, B., Prasetyo, E. N., Nyanhongo, G. S. and Guebitz, G. M. (2013). Antimicrobial enzymes: an emerging strategy to fight microbes and microbial biofilms. *Biotechnology journal*, 97-109.

Thomas, V. C., Thurlow, L. R., Boyle, D. and Hancock, L. E. (2008). Regulation of autolysis-dependent extracellular DNA release by *Enterococcus faecalis* extracellular proteases influences biofilm development. *Journal of bacteriology*, 5690-5698.

Torres Dominguez, E., Nguyen, P. H., Hunt, H. K. and Mustapha, A. (2019). Antimicrobial coatings for food contact surfaces: Legal framework, mechanical properties, and potential applications. *Comprehensive reviews in food science and food safety*, 1825-1858.

Totsika, M., Kostakioti, M., Hannan, T. J., Upton, M., Beatson, S. A., Janetka, J. W., Hultgren, S. J. and Schembri, M. A. (2013). A FimH inhibitor prevents acute bladder infection and treats chronic cystitis caused by multidrug-resistant uropathogenic *Escherichia coli* ST131. *The Journal of infectious diseases*, 921-928.

Tufail, M. S. and Liaqat, I. (2021). 2. Silver nanoparticles and their applications-A comprehensive review. *Pure and Applied Biology (PAB)*, 315-330.

Tufail, M.S., Liaqat, I., Andleeb, S., Naseem, S., Zafar, U., Sadiqa, A., Liaqat, I., Ali, N.M., Bibi, A., Arshad, N. and Saleem, G. (2022). Biogenic Synthesis, Characterization and Antibacterial Properties of Silver Nanoparticles against Human Pathogens. *Journal of Oleo Science*, 21291.

Uematsu, N. and Matsuzaki, K. (2000). Polar angle as a determinant of amphipathic α-helix-lipid interactions: a model peptide study. *Biophysical journal*, 2075-2083.

Ulm, H., Wilmes, M., Shai, Y. and Sahl, H. G. (2012). Antimicrobial host defensins–specific antibiotic activities and innate defense modulation. *Frontiers in immunology*, 249.

van Hoogevest, P., Liu, X. and Fahr, A. (2011). Drug delivery strategies for poorly water-soluble drugs: the industrial perspective. *Expert opinion on drug delivery*, 1481-1500.

Ventola, C. L. (2015). The antibiotic resistance crisis: part 2: management strategies and new agents. *Pharmacy and Therapeutics*, 344.

Vijayakumar, S., Vaseeharan, B., Malaikozhundan, B., Gobi, N., Ravichandran, S., Karthi, S., Ashokkumar, B. and Sivakumar, N. (2017). A novel antimicrobial therapy for the control of *Aeromonas hydrophila* infection in aquaculture using marine polysaccharide coated gold nanoparticle. *Microbial pathogenesis*, 140-151.

Vogel, A. and May, O. (2019). *Industrial enzyme applications*. John Wiley & Sons.

Waghu, F. H., Barai, R. S., Gurung, P. and Idicula-Thomas, S. (2016). CAMPR3: a database on sequences, structures and signatures of antimicrobial peptides. *Nucleic acids research*, D1094-D1097.

Wang, G. (2015). Improved methods for classification, prediction, and design of antimicrobial peptides. In *Computational Peptidology* (pp. 43-66). Springer.

Wang, G. (2020). The antimicrobial peptide database provides a platform for decoding the design principles of naturally occurring antimicrobial peptides. *Protein Science*, 8-18.

Wimley, W. C. (2010). Describing the mechanism of antimicrobial peptide action with the interfacial activity model. *ACS chemical biology*, 905-917.

Worthington, R. J., Richards, J. J. and Melander, C. (2012). Small molecule control of bacterial biofilms. *Organic & biomolecular chemistry*, 7457-7474.

Yasir, M., Willcox, M. D. P. and Dutta, D. (2018). Action of antimicrobial peptides against bacterial biofilms. *Materials*, 2468.

Yeaman, M. R. and Yount, N. Y. (2003). Mechanisms of antimicrobial peptide action and resistance. *Pharmacological reviews*, 27-55.

Zasloff, M. (2002). Antimicrobial peptides of multicellular organisms. *Nature*, 389-395.

Zhao, X., Wu, H., Lu, H., Li, G. and Huang, Q. (2013). LAMP: a database linking antimicrobial peptides. *PLoS One*, e66557.

Zhou, C. (2014). *Anti-bacterial, Anti-inflammatory and Anti-adhesive Coatings for Urinary Catherers* Drexel University].

Zhou, J. W., Luo, H. Z., Jiang, H., Jian, T. K., Chen, Z. Q. and Jia, A. Q. (2018). Hordenine: a novel quorum sensing inhibitor and antibiofilm agent against *Pseudomonas aeruginosa*. *Journal of agricultural and food chemistry*, 1620-1628.

Chapter 6

Insect-Derived Antimicrobial Peptides: A Novel Strategy for Biofilm Inhibition

Waiza Ansar[1] and Iram Liaqat[1*]
[1]Department of Zoology, Government College University, Lahore, Pakistan

Abstract

A biofilm is a multiplex layer of surface-linked microbial communities that makes the microbes more resistant to antibiotics and protects them from the immune response of the host. The ubiquitous nature of biofilms makes elimination very difficult. Nearly all biofilms, especially of highly resistant pathogens such as *Pseudomonas aeruginosa*, *Acinetobacter baumannii, Klebsiella pneumonia, Staphylococcus aureus*, *Enterobacter* spp., and *Enterococcus faecium,* cannot be eradicated with currently available drugs. A new antibiofilm strategy is to use insect-derived antimicrobial peptides (AMPs). In response to bacterial infections, insects produce AMPs as an immune response in their hemolymph. These peptides possess antibiofilm activity through different mechanisms to antibiotics, so the microbes may not become resistant against them. Instead of using AMPs alone, synergistic use of AMPs with antibiotics has shown an improved ability to inhibit biofilm formation. Various classes of insect-derived AMPs are known to have anti-biofilm effects, in particular, cecropin A against uropathogenic biofilms of *E. coli*.

Keywords: antimicrobial peptides, biofilm formation, immune response, resistant pathogens

*Corresponding Author's Email: iramliaq@hotmail.com.

In: Understanding Antibiofilm Activity
Editor: Iram Liaqat
ISBN: 978-1-68507-927-7

Introduction

Microbes have a remarkable ability to survive under almost all conditions, and they use sophisticated mechanisms to protect themselves from external agents. Bacteria exist either in planktonic (free-swimming) mode or in biofilms attached to surfaces (Watnick and Kolter, 2000).

A biofilm can be a visible slimy layer of organic matter in which bacteria or fungi are attached to solid biotic or abiotic surfaces (Sahoo et al., 2021). Biofilms are characterized by the presence of an extracellular matrix (ECM) that contains proteins, lipids, polysaccharides, glycolipids, and extracellular DNA (eDNA). One of the critical properties of the ECM is its ability to inactivate antimicrobial agents. Communities of microbial cells enclose themselves in these self-produced extracellular matrices and communicate with each other (Armbruster and Parsek, 2018). It is challenging to eradicate a biofilm as it makes the bacteria resistant to antibiotics. Biofilm-producing bacteria showed up to 1,000 times more resistance to antimicrobial agents than the same species of bacteria in planktonic mode (Sahoo et al., 2021). It is estimated that 80% of all bacterial infections are due to biofilms. In addition, biofilms evade the host defense mechanisms (Davies, 2003). It is challenging to eradicate biofilm with existing antibiotics. Pathogens, such as *E. faecium*, *S. aureus*, *K. pneumoniae*, *A. baumannii*, *P. aeruginosa,* and *Enterobacter* spp., are considered multiple drug/antibiotic-resistant and play a crucial role in the growth of nosocomial infections (Sahoo et al., 2021). The World Health Organization (WHO) states that there is an urgent need to develop new therapeutic agents against 12 bacterial pathogens. These pathogens are categorized into critical, high, and medium groups. Gram-negative species such as *P. aeruginosa* and *A. baumannii* are included in the critical priority group. Several Gram-positive species, such as methicillin-resistant *S. aureus* (MRSA) and vancomycin-intermediate are included in the high priority group (Shrivastava et al., 2018).

Antibiofilm agents must be able to target the lifestyle of biofilm. First, their killing ability must be rapid to target the cells before the biofilm community has fully developed and target the differences in the growth rates of cells. Cells located in biofilm periphery have direct contact with nutrients, oxygen and grow rapidly. Cells located in the deepest layer of biofilm face anoxia, lack nutrients and are slowly growing cells. So, metabolic heterogeneity is generated between these rapidly and slowly growing cells. Under stress conditions, it is possible to find quiescent, dormant, and persistent cells in the deepest layer of biofilm that play a crucial role in resistance to multiple drugs or antibiotics. Another feature of antibiofilm agents is their ability to interfere with ECM and penetrate biofilm architecture. They may also be able to interfere with the communication between cells.

Biofilm formation occurs through different phases: adhesion, maturation, and dispersal. Various approaches can inhibit biofilm formation compared to treating a mature biofilm. Some small molecules, such as proteins, peptides, and other organic molecules, have an antibiofilm ability. They must have particular characteristics that can inhibit biofilm through different mechanisms (Di Somma et al., 2020). The first line of defense of an organism against pathogens is the activation of innate immunity, followed by acquired immune responses consisting of T and B cells that target the specific antigens. Along with acquired immune mechanisms, peptides are produced that provide defense against these pathogens (Reddy et al., 2004).

Peptides are biomolecules, basic in nature, and found in all organisms. Most commonly, proteins are broken down by chemical degradation and enzymatic cleavage to yield peptides of linear chains of 2 -20 amino acids. These peptides exhibit important properties like antimicrobial, anticancer, virucidal, fungicidal, antioxidative, and immunomodulatory effects. Peptides derived from insects offer a novel therapeutic approach to combat biofilms. This group of molecules, termed 'antimicrobial peptides'(AMPs), act as an immune defense mechanism (Bals, 2000). Their potential to overcome bacterial resistance make them promising candidates for therapeutic drugs. Antimicrobial peptides may be especially abundant in certain phagocytes and can be obtained from insects' reproductive tract and fat bodies. The expression of antimicrobial peptides can be constitutive, inducible, or both (Lehrer and Ganz, 1999). Insects are highly resistant to bacterial infections, and upon pathogenic invasion, a genetic cascade is activated, leading to the synthesis of antibacterial peptides released into the hemolymph. In insects, AMPs are divided into four main groups based on structure and amino acid composition (Otvos, 2000). These groups include proline-rich peptides, cysteine-rich peptides, alpha-helical peptides, and glycine-rich proteins (Bulet and Stocklin, 2005). Defensins kill Gram-positive strains selectively. Proline-rich and glycine-rich peptides are predominantly active against Gram-negative strains, while the cecropins are active against both types.

Bacterial Biofilm Formation and Its Composition

Building a biofilm is a highly complex process, as it requires a sequence of distinct and highly well-organized steps. In phase 1, the bacterium chooses the substrate for attachment. The characteristics of cell surface and substratum significantly affect microbial attachment. Substrata can range from hydrophobic to hydrophilic materials. Biofilms on hydrophobic or rougher materials grow rapidly. Characteristics of cell surfaces like flagella, fimbriae, and pili enable the cells to remain attached to the surface by overcoming any repulsive forces (Donlan, 2001). After attachment, they form a temporary relationship with neighboring microbes attached to that surface. A film is formed by a mass of dissolved organic molecules like proteins, glycolipids, glycopeptides, and extracellular DNA (eDNA) on the solid: water interphase because of the hydrophobic nature of these molecules. Then bacteria form small groups, which are known as micro-colonies. Just as tissue is the basic unit for the growth of more complex organisms, similarly, the microcolony is the basic structural unit of the biofilm. Most of the microcolony is comprised of EPS matrix (75-90%), and the rest is cells (10–25%), depending on the type and number of species. This extracellular material mediates the microbes' interaction and protects them against external stress or antibiotics (Sahoo et al., 2021). After biofilm maturation, bacterial colonies express different genes and presume a discrete phenotype to their planktonic counterparts (Stoodley et al., 2002). During biofilm formation, some species of bacteria interact with each other by a signaling pathway called quorum sensing (Jamal et al., 2018; Liaqat et al., 2014). As a result of cell division and growth, bacterial aggregates separate from the biofilm and cause systemic infections (Donlan, 2001). Biofilm formation can occur on a variety of surfaces comprised of living tissues as well as prosthetic implants. About 99% of the microbial population exists in the form of biofilms containing diverse microorganisms, and high antibiotic resistance is observed (Garrett et al., 2008). For example, biofilms in the oral cavity contain more than 500 microbes. A developed biofilm consists of groups of microorganisms, often described as being shaped like mushrooms, separated by spaces that are filled with the surrounding fluid (Tenke et al., 2006).

Impact of Bacterial Biofilm on Human Health

The increasing number of patients infected with antibiotic-resistant bacteria is one of the greatest challenges to human health worldwide, as declared by the WHO (Zhang et al., 2018). About 99% of microbes exist in biofilm mode. So, it is recognized that persistent infection in cystic fibrosis (CF) patients (Høiby, 2017) and pathogenesis in chronic infections (Bjarnsholt, 2013) are the results of biofilm involvement. Biofilms cause persistent infections in several parts of the body, such as the urinary tract, teeth, heart valves, middle ears, lungs, prostate, bones, and nasal passages. Biofilm structures have been observed in the upper surface of respiratory epithelia in CF patients and lungs lavage samples. Biofilm-producing pathogens have also been isolated from rhinosinusitis, pharyngitis or sore throat, and whooping cough (Vestby et al., 2020). In most cases, the presence of a biofilm increases the complications of the disease.

Bacterial prostatitis (BP) is a urinary tract infection (UTI) that can involve a number of pathogenic bacteria (i.e., *P. aeruginosa*, *Klebsiella* spp., *E. coli*, *P. mirabilis*, *Enterobacteriaceae*, and *E. faecalis)* that cause bladder infection, bacteremia, infertility, urosepsis, prostatic abscesses, kidney damage, septicemia, and death. Complications of UTI are enhanced by *E. coli* (biofilm-producing bacteria), which commonly occur in UTIs (Flores-Mireles et al., 2015). Pathogenic bacteria also affect the integumentary system of humans. Damaged living tissues acquire severe infections by biofilm-producing or planktonic bacteria. The cases become more complicated in the case of immunocompromised patients. Biofilms are the basic cause of transforming wounds into chronic stages and also cause delays in wound healing (Bjarnsholt et al., 2011).

Bacterial vaginosis (BV) is the most common vaginitides in women of reproductive age. It is caused by a decrease in protective *Lactobacillus* spp., due to biofilm formation by anaerobic bacteria (*Gardnerella vaginalis* and *Atopobium vagnae)* (Sobel, 2000).

Typhoid fever is a human-specific food-borne disease caused by *Salmonella enterica* serovar *Typhi*. It is estimated that 3 to 5% of people infected with typhoid fever become chronic carriers. The mechanism of chronic carriage involves biofilm formation on the surfaces of gallstones. Antibiotics that are generally effective against acute infection are ineffective against the biofilm-associated chronic colonization of the gall bladder (Neiger et al., 2019).

Previously, it was considered that planktonic bacteria are the cause of chronic wound infections. But more recently, it has been found that it is the coverage of the wound surface by biofilms that contribute to inflammation and wound healing failure (Percival et al., 2012). About 60% of chronic wound infections are due to biofilms (James et al., 2008).

Microorganisms such as species of *Streptococci*, *Enterococci*, and *Staphylococci* isolated from endocarditis all have the ability to form biofilms and resist conventional antibiotic therapies. It is reported through fluorescence *in situ* hybridization that biofilm-producing microorganisms are also found in atherosclerotic arteries, where they enhance the pathogenicity (Vestby et al., 2020).

An emerging field of research is thus the development of new therapeutic options to combat biofilms and understand the mechanism of their antibiotic resistance (Ciofu et al., 2015).

Bacterial Virulence Factors Involved in Biofilm Formation

Bacterial cells in a biofilm are more resistant to antibiotics than planktonic cells. Drug resistance factors are well understood in planktonic cells but are more complex for cells in biofilms. Understanding of these resistance mechanisms is crucial for developing potent antibiofilm agents (Munita and Arias, 2016).

Virulence factors involved in the regulation of biofilm formation are laminin-binding proteins, elastin binding proteins, fibronectin-binding protein A, biofilm-associated proteins, collagen-binding proteins, and serine aspartate repeat proteins (Atshan et al., 2012). In biofilm formation, the crucial step is bacterial cell attachment to the host. In the case of *Enterococcus* spp. a surface protein is involved in cell adhesion, colonization and persistence in the urinary tract. With the help of collagen-binding proteins and aggregation proteins, they evade the host immune responses. Some other virulence factors, such as secretory antigens, surface anchor proteins, and endocarditis biofilm-associated pili, also mediate the biofilm formation. Most virulence genes such as autolysis, cytolysin A, hyaluronidase, and serine protease regulate host cell autolysis.

Another Gram-pathogen is *Streptococcus pneumoniae* which contains the virulence factor polysaccharide intercellular adhesin (PIA). This factor is responsible for matrix formation in *S. aureus* biofilms. In the absence of PIA, eDNA and proteins support biofilm formation.

A Gram-negative bacterium, *K. pneumoniae,* is commonly found in the intestinal tracts of humans (Piperaki et al., 2017). It is the leading cause of pneumonia in alcoholics and diabetic patients before the period of antibiotics. But now this bacterium is mainly associated with hospital-acquired infections. The formation of biofilm in catheter-linked infections is promoted by type-3 fimbriae caused by *K. pneumoniae* (Paczosa and Mecsas, 2016).

Some virulence factors that increase the pathogenicity of *P. aeruginosa* (Gram-negative bacterium) are pili type IV, flagella, lipopolysaccharides, exotoxin A, alginate, protease enzyme, and type 3 secretion system (T3SS). Pilli type IV is involved in bacterial cell adhesion to host cells, and lipopolysaccharides activate both host innate and adaptive immune responses. T3SS causes phagocytosis of host cells by promoting the transport of proteins from the cytoplasm of *P. aeruginosa* to the cytosol of host cells.

Enterobacteriaceae are a family of Gram-negative bacteria commonly found in the intestinal tract and in the urinary tract, where they cause urinary tract infections (UTIs) (Magiorakos et al., 2017). Virulence factors that promote the development of these biofilms are serum resistance, adhesins, and hemolytic production (Rossi et al., 2018).

Another Gram-negative bacterium, *A. baumannii,* causes chronic infections in humans, mainly by biofilm formation on biotic and abiotic. Virulence factors that mediate this biofilm formation are proteins involved in biofilm maturation, while outer membrane protein-A targets the epithelial cells. A two-component system, BfmS and BfmR are also contributes equally in biofilm formation.

Biofilm Destruction Strategies

The development of antimicrobial agents for biofilm eradication is an emerging area of research. Several antibiofilm agents have already been developed such as antimicrobial lipids (e.g., docosahexaenoic acid and glycerol monolaurate) (Schlievert and Peterson, 2012). Quaternary ammonium compounds (e.g., tris-QAC-10, XF73) (Forman et al., 2016). Redox-active secondary metabolites, nitric oxide-releasing antibiotics (ciprofloxacin – nitroxides 27), and AMPs (e.g., G10, LL-37, novispirin etc.). Overall, AMP use is considered the best therapeutic approach to counteract biofilm-associated infections because of their distinctive characteristics.

Antimicrobial Peptides (AMP)

Naturally, almost all living organisms produce AMPs as an immune response in their body. These proteins have the potential to kill fungi, bacteria, yeasts, viruses, and even cancer cells. The first AMP was discovered in 1939, followed by the discovery of some essential antimicrobial peptides in the 1980s. Several AMPs were synthesized in the laboratory artificially by mimicking the original sequence or with the help of computer design. No matter the origin, almost all peptides share some common properties, such as a net positive charge and small size (15-30 amino acids). They all have amphipathic activity, and in some cases, they are also membrane activators. Positively charged peptides are attracted mainly by cell membranes of bacteria and biofilm sites. Peptides are classified based on structure, alpha-helical, beta-sheet, and extended coil. Several peptides contain both alpha-helix and beta-sheet domains (Figure 1). In alpha-helical peptides, most commonly, amide groups are present at the C-terminal, which enhances the antimicrobial activity as it increases the electrostatic interaction between the positively charged peptides and negatively charged membranes (Mura et al., 2016). Almost all β sheet antimicrobial peptides contain cysteine residues which form disulfide bonds.

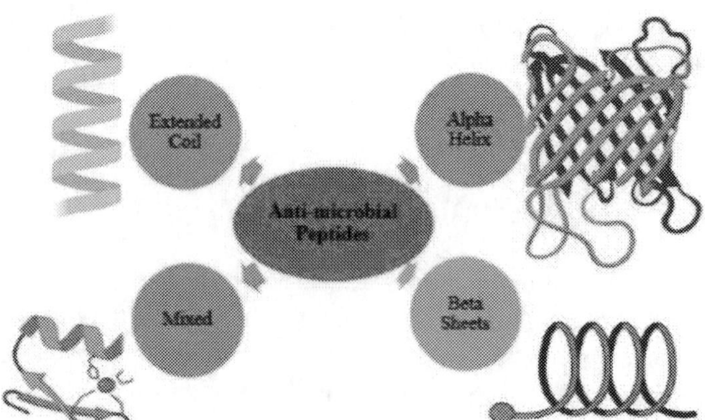

Figure 1. Antimicrobial peptides are classified into four main types i.e., extended coil, alpha-helix, beta-sheet, and mixed (both alpha-helix and beta sheets) based on their structure.

AMPs are produced as antibiotics by plants and insects against microbes. But microbes also secure their environmental niche by AMP production (Zhang and Gallo, 2016). Bacteria do not defend themselves from other bacterial species by producing AMPs, but they target other species of bacteria by reducing competition for nutrients. AMPs target the slowly growing bacteria in biofilms and quickly kill them (Tiwari et al., 2015). Bacteria living in the biofilm mode are resistant to antibiotics due to their slow growth rate and low metabolic rate in such communities. AMPs act on slowly growing and even non-growing cells because they can permeabilize the bacterial membranes and cause pore formation in cytoplasmic membranes (Batoni et al., 2011). In addition to natural peptides, synthetic peptides, such as a derivative of LL -37 AMP, also act as antibiofilm agents (Bahar and Ren, 2013).

Antimicrobial Peptides from Insects

Insects are the source of a wide variety of AMPs. Insect AMPs have proven beneficial in the agricultural, pharmaceutical, and medical fields. AMPs are produced from the insect's fat body, secreted into the hemolymph, where they diffuse around the entire body. Various epithelia can also synthesize them. In insects, antibacterial activity was first observed in the pupae of the giant silk moths *Samia cynthia* and *Hyalophora cecropia.* Cecropin was the first insect AMP purified from *H. cecropia* in 1980. Later, more than 150 AMPs were identified. These AMPs are of small size, basic in nature, and show antibacterial activity against bacteria and fungi. Some also target viruses and parasites. These compounds are effective against both Gram-positive and Gram-negative bacteria, and some of them can prevent biofilm formation in highly resistant pathogens (i.e., *E. coli*). One of the potential agents, defencin-1 is isolated from *Apis melifera* (honey bee) and is responsible for the destruction of the planktonic and biofilm cells of *Streptococcus agalactiae, P. aeruginosa, S. aureus,* and *E. faecalis* (Sojka et al., 2016). Another bioactive peptide, melittin, also isolated from *A. mellifera* possesses antibiofilm properties against the pathogenic bacteria *P. aeruginosa, E. coli* and *K. pneumoniae* (Danihlík et al., 2015). Another antimicrobial peptide, coprisin, has shown the potential to inhibit biofilms (Memariani et al., 2018). Cecropins, derived from *H. cecropia,* acts by lysing bacterial membranes and inhibiting proline uptake (Bechinger and Lohner, 2006). Cecropin B was found to be effective against *E. coli*. Mastoparan-1, derived from *Polybia paulista,* has the potential to fight against methicillin-resistant *S. aureus* (Memariani et al., 2018).

However, AMPs originally derived from insects can have limitations. For example, they can affect the human neutrophil elastase recruited during infections (Romoli et al., 2019). Bacterial resistance to AMPs is rare because AMPs are attracted to their negatively charged lipid bilayer membranes (Bahar and Ren, 2013).

Mechanisms of Antibiofilm Peptides against Pathogens

In insects, antimicrobial proteins and peptides are produced in response to bacterial infection as an immune response. These peptides are small and contain cationic groups. The structure of many peptides is observable only through membrane interaction (Jenssen et al., 2006). For instance, indolicidin (a neutrophil peptide) is unstructured in an aqueous medium and adopts a boat-like structure after interacting with membranes. The antimicrobial activity of each peptide is affected by its composition, character, and physiological condition. A peptide contains cationic residues in large numbers that statistically interact with negatively charged molecules of the cell surface. These cationic peptides can potentially kill the cell directly or by immunomodulatory action (Mukhopadhyay et al., 2020). Due to the difference in cell wall composition, peptides are more effective against Gram-negative than Gram-positive bacteria. Nonspecific mechanisms of inhibition target the oxidative pathways, membranes, and intracellular molecules (Band and Weiss, 2015). The potential of AMPs as therapeutic agents compared to conventional antibiotics depends on their ability to kill bacterial cells without causing any cytotoxicity to mammalian cells. Various peptides have different modes of action according to their structures (Gomes et al., 2018).

The most accepted models that define the bacterial membrane disruption of AMPs mechanisms are Barrel-stave, Toroidal-pore, and Carpet (Zasloff, 2019). According to the barrel stave mechanism, transmembrane pore formation occurs in sequential steps. First, monomers of peptides interact with the cell membrane in an alpha-helical fashion. Bundles of monomers localize on the cell membrane, followed by the insertion of these monomers into the hydrophobic core of the membrane. Multiple helices are recruited into the core and make the pore larger, causing leakage of cellular content and finally, cell death occurs (Reddy et al., 2004). The carpet model explains membrane disruption in the following steps. First, peptides bind to phospholipids' head groups of the membrane, followed by peptides alignment so that their hydrophilic surfaces face the hydrophobic head groups of the membrane. Further peptides reorient themselves and enter the cytoplasmic core. They disrupt the curvature of the bilayer, and disintegration of the membrane occurs (Matsuzaki, 2019). According to the toroidal pore model, peptides in helical configuration initially interact and attach to bacterial membrane. By lining the water core with both lipid head groups and peptides, they cause bending of the lipid bilayer and pore formation. Commonly, the AMP, melittin, induces this kind of transmembrane pore in bacterial membranes (Brogden, 2005).

Peptides kill or inhibit the growth of microbes through the steps of attraction, attachment, and insertion. Generally, AMPs target the anionic membranes of bacteria (Hollmann et al., 2018). Initially, peptides interact with microbial surfaces. They penetrate the complete depth of polysaccharides on the bacterial surface, interact with the lipopolysaccharides on the outer membrane, and disrupt its structure. In the case of Gram-positive bacteria, peptides interact with anionic teichoic acid, gather on the surface and gain access to the inside of cytoplasmic membranes (Zasloff, 2019).

The formation of transmembrane pores is not the only mechanism of microbial killing. Translocated antimicrobial peptides can change the ingrowth of the cytoplasmic membrane (septum formation), inhibit the synthesis of nucleic acid, cell wall, and proteins, and inhibit the activity of enzymes. Microorganisms are rapidly killed by all of these mechanisms (Brogden, 2005).

Classification of Insect AMPS

Insect AMPs can be classified into four families based on certain structures or sequences.

1. Alpha-helical peptides deprived of cysteine residues (cecropin and moricin),
2. Cysteine-rich peptides (insect defensin and drosomycin),
3. Linear extended peptides
    - proline-rich peptides (apidaecin, drosocin, and lebocin) and
    - glycine-rich peptides (attacin and gloverin) (Bulet and Stocklin, 2005).

Cysteine-rich peptides are the major constituents of innate immunity due to their molecular action in inhibiting biofilms in various organisms, including insects. Besides these, new peptides such as diptericins, drosomycin metchnikowin, and ponericins are still being investigated. Glycine and proline-rich peptides are primarily effective against Gram-negative bacteria (Wu et al., 2018).

Alpha-Helical Peptides

Cecropin

Insects synthesize cecropins in their fat body cells and hemocytes (Kim et al., 2010). Cecropin (Cec) peptides were first isolated from the pupae of *H. cecropia* (cecropia moth) and contain no cysteine residues. Later, these peptides were also identified in two orders, Coleoptera and Diptera. Peptides similar to Cec have also been identified in *Helicobacter pylori* (bacterium) (Bylund et al., 2001). Cecropins form alpha-helices (Otvos, 2000) are categorized into six groups, Cec A, B, C, D, E, and F. Cecropin D is the major cecropin, containing 37 amino acids. These peptides are active against fungi, Gram-positive and Gram-negative bacteria (Vizioli et al., 2000). Cecropins are amidated at the C-terminal end to interact with liposomes, resulting in antibacterial activity.

Role of CecA in the Disruption of Uropathogenic *E. coli* Biofilms

Uropathogenic *E. coli* causes almost 90% of urinary tract infections in humans (UPEC) (Flores-Mireles et al., 2015). Treatment becomes difficult because of their biofilm-forming ability on epithelial cells of the bladder and urinary catheters (Toval et al., 2014). These bacteria resist current antibiotics by quorum sensing (Liaqat et al., 2014), and an ECM is resistant to permeation by the antibiotic, efflux pumps, and persisted cells (Koo et al., 2017). Antibiofilm agents must have the properties to overcome all these strategies to inhibit biofilm and kill planktonic cells (Liaqat et al., 2022).

Insect AMP, CecA, isolated from *Galleria mellonella* (the greater wax moth), has the potential to eradicate biofilm formed by UPEC. The combined use of CecA with nalidixic acid inhibits bacterial eDNA release, surface attachment, and energy-dependent MDR efflux pumps. CecA also binds with lipopolysaccharides (LPS) and makes the membrane permeable for the uptake of nalidixic acid. After diffusing nalidixic acid into the cell, they disrupt the biofilm of UPEC by binding with different molecular targets such as DNA and RNA. At 50µg/mL minimum inhibitory concentration of CecA and 0.5ng/mL of nalidixic acid, synergistic activity is indicated with no cytotoxicity in fibroblasts and erythrocytes (Fenner, 2020).

Moricin

Moricin is a novel antibacterial peptide derived from *Bombyx mori* (the silkworm). It is a basic peptide with 42 amino acids and possesses vigorous antibacterial activity against Gram-positive and Gram-negative bacteria (Hara and Yamakawa, 1996). This peptide most significantly targets *S. aureus*. Moricin acts by targeting bacterial cytoplasmic membranes. A predicted α-helix present at the N-terminal portion of the peptide enhances the membrane permeability (Hara and Yamakawa, 1995). Moricin consists of five basic amino acid residues in the C-terminal, and its amino acid sequence has no significant similarity with other peptides (Hemmi et al., 2002).

Cysteine-Rich Peptides

Defensins

Defensins are small (36 - 46 amino acids) antimicrobial peptides with six cysteine residues and three disulfide bridges (Otvos, 2000). They mainly kill Gram-positive bacteria, including *S. aureus*. Defensins have the unusual feature of selectively killing Gram-positive bacteria (Otvos, 2000) Defensins are much less responsive against Gram-negative bacteria and yeasts. Insect hemolymph, intestinal epithelial surfaces, trachea, and the granules of phagocytes contain defensin peptides. Defensins are found in both endopterygote (i.e., Diptera, Trycoptera, Coleoptera, and Hymenoptera) and exopterygote insects (i.e., Hemiptera and Odonata) (Yi et al., 2014). These are classified according to structure into classical defensins, β-defensins, and insect defensins. The AMP lucifensin (Lucilia defensin) defensin analog, isolated from *Lucilia sericata* (green bottle fly), consists of 44 amino acid residues and protects the maggots against pathogenic elements of microorganisms present in the wound (Čeřovský et al., 2010). Defensins spontaneously kill bacteria; for example, one-minute contact of 0.5mM defensin is enough to kill *M. luteus*. The insect defensin family includes lucifensin, royalisin, sapecins, coprisin, spodoptericin, phormicins, and tenecin-1 (Yi et al., 2014).

Table 1. Antimicrobial activity of defensins and their derivatives

Defensin	Species	Activity	References
Lucifensin	*L. sericata*	Gram-positive	(Čeřovský et al., 2010)
Royalisin	*A. mellifera*	Gram-positive	(Bílikova et al., 2015)
Sapecin B	*Sarcophaga peregrina*	Gram-positive and Gram-negative	(Manabe and Kawasaki, 2017)
Coprisin	*Copris tripartitus*	Gram-positive and Gram-negative	(Hwang et al., 2009).
Defensin-1	*Acalolepta luxuriosa*	Gram-positive and Gram-negative	(Ueda et al., 2005)
Spodoptericin	*Spodoptera frugiperda*	N/A	(Volkoff et al., 2003).
Gallerimycin	*Galleria mellonella*	Fungi	(Schuhmann et al., 2003)
*Spli*Def	*Spodoptera littoralis*	Gram-positive and Gram-negative	(Seufi et al., 2011)
Termicin	*Pseudacanthotermes spiniger*	Fungi	(Da Silva et al., 2003)

Drosomycin

Drosomycin is an antifungal peptide and exhibits no antibacterial activity. It belongs to the family of inducible peptides (Fehlbaum et al., 1994). As its name indicates, this peptide consisting of forty-four amino acid residues, extracted from *Drosophila melanogaster*. Its three-dimensional structure involves alpha-helix and three-stranded twisted beta-sheets stabilized by three disulfide bridges. It has potent effects againstphytopathogenic filamentous fungi (Zhang and Zhu, 2009).

Glycine Rich Peptides

Attacin and gloverin are the two crucial glycine-rich insect-derived peptides.

Attacin

Attacin is an immune protein isolated from *H. cecropia* (the giant silk moth), and its different isoforms were purified from hemolymph after immunizing the pupae with the bacterium *Enterobacter cloacae*. These isolated isoforms can be classified into two groups: the basic group, attacins from A to D, and the acidic group from E to F (Buonocore et al., 2021). Amino acid sequences of these two groups are similar but the acidic group contains more Asp residues than the basic group. Two different genes encoded both groups. A hybrid of two proteins, attacin, and coleoptericin is obtained by using three Glu–Ala–Ala–Ala–Lys peptide sequences as a helical linker and shows more significant antibacterial activity against *E. coli*, *Bacillus subtilis*, and *Burkholderia glumae* than the parental proteins (Lee et al., 2013). Attacin acts by blocking the major protein synthesis on the outer membrane of Gram-negative bacteria. It inhibits the synthesis of outer membrane proteins of *E. coli* at the transcriptional level without entering the cytoplasm. In this way, it disrupts the stability of the outer cell wall and allows the bacterial cells to grow in longer chains (Wu et al., 2018). The humoral immune system of *H. cecropia* contains attacins as antibacterial proteins, along with cecropin and lysozyme. Many reports are available about the antibacterial activity of attacin against Gram-negative bacteria, but no data is available regarding antibiofilm activity.

Gloverin

Gloverin is a glycine-rich antimicrobial peptide isolated from two species belonging to Lepidopteran, *Hyalophora gloveri* (the giant silkmoth) and *Helicoverpa armigera* (the old world bollworm) (Badapanda and Chikara, 2016). Gloverin consists of 18% glycine residues. Amino acid sequences of this peptide have no similarity with any other AMP. They possess intense antibacterial activity against Gram-negative bacteria but are not effective against Gram-positive bacteria, yeasts, pestivirus, and mammalian cell lines. Gloverin acts by blocking the synthesis of bacterial outer membrane proteins and hence increases the membrane permeability of Gram-negative bacteria. Its mode of action is most similar to attacin (Bartlett et al., 2002).

Proline-Rich Peptides

These are a group of cationic peptides containing many proline residues. Their mode of action is different from other antimicrobial peptides, as they do not lyse the bacterial membranes (Scocchi et al., 2011). They bind to the 70S ribosome of bacteria and inhibit the synthesis of bacterial proteins (Krizsan et al., 2014).

Drosocin

Drosocin isolated from *D. melanogaster* consists of 19 amino acids. Glycosylation (post-translational modification) is a vital modification by some proline-rich amino acids for maximum biological activity (Lele et al., 2015). Drosocin is O-glycosylated and inhibits the growth of fungi and *E. coli* (Imler and Bulet, 2005). The activity of drosocin is terminated by the deletion of the first five N terminal residues.

Lebocins

Lebocins are isolated from the hemolymph of *B. mori* (the silkworm) after immunization with *E. coli* (Kaneko et al., 2007). These peptides are O-glycosylated and consist of 32 amino acids. About 41% of the amino acid sequence of lebocin is similar to a sequence of abaecin produced by *A. mellifera* (Yi et al., 2014). Natural peptides can be modified in the laboratory to enhance their antimicrobial activity. For instance, analogs of apidaecin and drosocin can be glycosylated or unglycosylated. Synthesized glycosylated derivatives have less antimicrobial activity than unglycosylated containing intrachain, and disulfide bonds (Gobbo et al., 2002). Metchnikowin is another proline-rich AMP isolated from *D. melanogaster* that is active against Gram-positive bacteria but does not show antimicrobial activity against Gram-negative bacteria (Manniello et al., 2021). Apidaecins, isolated from the lymph fluid of the honeybee (*A. mellifera*), provide a defense to the honeybee against bacteria (Zhou et al., 2008).

Table 2. Insect's antimicrobial peptides derivatives have antibiofilm activity

Peptide	Insect source	Activity against highly resistant pathogens	Mechanism of action	References
Mastoparan- C	*Vespa crabro*	*S. aureus* and *P. aeruginosa*	Membrane disruption	(Chen et al., 2018).
A3 analog of AamAP1 (derived from scorpion venom)	*Androctonus amoerux*	*S. aureus*	Membrane disruption	(Almaaytah et al., 2018).
Defensin-1,	*A. mellifera*	*P. aeruginosa, S. aureus, E. faecalis* and *S. agalactiae*	Membrane disruption	(Sojka et al., 2016).
Mauriporin	*Androctonus mauritanicus*	Methicillin resistant *S. aureus*	Membrane disruption	(Almaaytah et al., 2014).
Mastoparan - 1	*P. paulista*	Methicillin-resistant *S. aureus*	Destabilization of bacterial membrane	(Memariani et al., 2019).
Coprisin	*C. tripartitus*	*E. coli, S. aureus, E. faecium, P. aeruginosa* and *Streptococcus mutans*	Membrane disruption	(Hwang et al., 2013).
Cecropin A	*G. mellonella*	Uropathogenic *E. coli*	Membrane disruption	(Kalsy et al., 2020)
Mellitin A	*A. mellifera*	*K. pneumoniae, E. coli, P. aeruginosa*	Membrane disruption	(Memariani et al., 2019)
BmKn–22	Venom of scorpion Mesobuthus martensii Karsch	*P. aeruginosa*	Cell signaling	(Teerapo et al., 2019).

Insect-derived AMPs that control bacterial biofilms include battacin, cecropin A and mellitin A, pleurocidin, bactenecin, and A3 analog of AamAP1 (derived from scorpion venom), and defensin-1 (De La Fuente-Núñez et al., 2016; Almaaytah et al., 2018; Krishnan et al., 2020).

Synergistic Study of AMPS with Antibiotics

Interest is growing in using a combination of AMPs in synergistic action with conventional antibiotics. This approach is promising as AMPs enhance the permeabilization of bacterial membranes, enabling antibiotics to enter the cell at higher concentrations (Hollmann et al., 2018). Bacteria can usually resist antibiotics by pumping them out of the cell through efflux pumps (Wu et al., 2018). As AMPs act on bacterial membranes by inducing pores and disturbing the osmotic balance, this is most likely the pathway by which the antibiotics overcome the efflux pumps and enter the cell. The combination of AMPs with antibiotics affects different targets inside the cell and thus improves the synergistic activity.

As AMPs activity depends on their structure (Gomes et al., 2018), different AMPs may be used in different circumstances and against various bacteria. Antibiofilm activity is recorded by the AMP CecA, isolated from *G. mellonella*. When CecA is used with nalidixic acid, it shows significant antibiofilm activity against UPEC infections at low concentrations (Kalsy et al., 2020). The antimicrobial activity of AMP 1037 against planktonic cells is fragile but it inhibits biofilm formation by *Burkholderia* spp. that are highly resistant to polymyxin B and other cationic peptides. However, this peptide does not affect the biofilm formation of *P. aeruginosa* PA2204, but triggers its cell proliferation (De La Fuente-Núñez et al., 2012). AMP LL-37 inhibits bacterial biofilm formation *in vitro* and affects the biofilms of existing, pre-grown *P. aeruginosa* (Overhage et al., 2008). The AMP, NA-CATH:ATRA1-ATRA1, inhibits biofilm formation in *S. aureus* (Dean et al., 2011). These AMPs inhibit the expression of biofilm-forming coded proteins in genes. Biofilm detection is also improved by binding AMPs to eDNA (Das et al., 2010). The biofilm-forming regulatory system of microorganisms plays essential role in resisting the effects of AMPs (Herbert et al., 2007). This resistance can be controlled by using a mixture of AMPs and antibiotics. The combination of AMPs with standard antibiotics is effective against biofilms formed by *P. aeruginosa* (Dosler and Karaaslan, 2014).

A combination of AMP A3 isolated from *A. amoerux* with conventional antibiotics shows the synergistic action against multi-drug resistant (MDR) bacteria. This combination may be a novel treatment strategy against biofilm-forming bacteria. The mixture of BmKn–22 derived from BmKn-2, isolated from the venom of the scorpion (*M. martensii* Karsch), with the antibiotic azithromycin, also prevents biofilm formation in *P. aeruginosa* (Teerapo et al., 2019). Coprisin is a peptide-like defensin derived from the dung beetle (*C. tripartitus*) that shows antibiofilm activity alone and in a mixture with antibiotics such as ampicillin, chloramphenicol, and vancomycin (Hwang et al., 2013). A combination of coprisin with ampicillin is effective against *S. mutans, P. aeruginosa,* and *E. faecium*. High synergistic activity is observed when using a combination of coprisin and vancomycin against all Gram-negative bacteria and *S. aureus*. Peptide melittin A in combination with doripenem is synergistically was observed to be effective against *A. baumannii* and *P. aeruginosa*. Also, MDP1 and MDP2, derived from melittin, showed activity against *E. coli* and *S. aureus* (Akbari et al., 2018). The antibiofilm peptide, 1018, blocks the signal ppGpp, which is vital in biofilm formation. It eradicates the mature biofilms of *E. coli*, methicillin-resistant *S. aureus*, *K. pneumoniae,* and *A. baumannii*. At low levels, it causes biofilm dispersal and at higher concentrations, causes the death of biofilm cells (De la Fuente-Núñez et al., 2014). A combination of AMPs with generic antibiotics is a suitable alternative to using antibiotics alone.

Table 3. Synergistic activity of insects derived AMPs with conventional antibiotics against the biofilms of highly resistant pathogens

Antimicrobial peptides	Antibiotics	Synergistic activity against biofilms of pathogens	References
Cecropin A	Nalidixic acid (NAL)	Uropathogenic *E. coli*	(Kalsy et al., 2020)
BmKn–22	Azithromycin	*P. aeruginosa*	(Teerapo et al., 2019)
Melittin	Doripenem	Multi drug resistant *P. aeruginosa* and *A. baumannii*	(Akbari et al., 2019)
Coprisin	Ampicillin	*P. aeruginosa, S. aureus* and *E. faecium*	(Hwang et al., 2013)
	Vancomycin	All Gram-negative bacteria including *S. aureus*	
	Chloramphenicol	*E. coli, S. mutans, S. aureus*	

Clinical Trials

In 2015–2019, approximately 208 drugs have been approved, of which 15 are peptide-based (de la Torre and Albericio, 2020). The development of peptide-based drugs is gaining more attention because of their advantages over antibiotics. Several AMPs are under clinical development for use against bacterial pathogens. Examples are pexiganan, omiganan, and LTX-109 from immune components of animals. However, while many AMPs show positive responses in clinical trials, they fail due to toxicity concerns (Koo and Seo, 2019).

Clinical success rates and high antibiofilm activity against resistant pathogens can be improved by specific modifications (De La Fuente-Núñez et al., 2016). Derivatives such as BmKn-22, A3, Pro10-1D, Defensin-1, and synthetic derivatives such as LL7-31, 1018, AS10, BMAP27-melittin and battacin are modified forms of parent AMPs designed to improve their antibiofilm activity against resistant pathogens.

Conclusion

It is apparent that AMPs are the main molecules that provide insects with a defense against microbes. Biofilms are a reason for persistent and chronic infections and make treatment with conventional antibiotics ineffective. New research is needed to find novel therapeutic agents against biofilm-associated diseases. AMPs could be one such option and show diverse mechanisms to kill microbes. It is not easy for microbes to become resistant to AMPs in comparison to conventional antibiotics. AMP's charge, sequences, hydrophobicity, and amphipathicity are important in this respect. In aprticular, AMPs derived from insects have the potential to kill planktonic as well as biofilm-producing pathogens. They can be used to target antibiotic-resistant pathogens i.e., multidrug-resistant *S. aureus*, uropathogenic *E. coli,* and carbapenem-resistant *A. baumannii*. Combined use of AMPs with antibiotics may be the best alternative to the use of antibiotics alone. While many peptides, isolated from insects, show antimicrobial activity against planktonic bacteria, only a few isolated peptides and their derivatives have antibiofilm potential. Yet, no insect-derived AMP has been progressed to clinical trials. Further research on AMPs is needed in different fields to discover useful AMPs. Research should be focused on screening all potential peptides isolated from insects for antibiofilm activity.

References

Akbari, R., Hakemi-Vala, M., Pashaie, F., Bevalian, P., Hashemi, A., and Pooshang Bagheri, K. (2019). Highly synergistic effects of melittin with conventional antibiotics against multidrug-resistant isolates of *Acinetobacter baumannii* and *Pseudomonas aeruginosa. Microbial Drug Resistance*, 193-202.

Akbari, R., Vala, M. H., Hashemi, A., Aghazadeh, H., Sabatier, J.-M. and Bagheri, K. P. (2018). Action mechanism of melittin-derived antimicrobial peptides, MDP1 and MDP2, de novo designed against multidrug resistant bacteria. *Amino acids*, 1231-1243.

Almaaytah, A., Farajallah, A., Abualhaijaa, A. and Al-Balas, Q. (2018). A3, a scorpion venom derived peptide analogue with potent antimicrobial and potential antibiofilm activity against clinical isolates of multi-drug resistant gram positive bacteria. *Molecules*, 1603.

Almaaytah, A., Tarazi, S., Alsheyab, F., Al-Balas, Q. and Mukattash, T. (2014). Antimicrobial and antibiofilm activity of mauriporin, a multifunctional scorpion venom peptide. *International Journal of Peptide Research and Therapeutics*, 397-408.

Armbruster, C. R. and Parsek, M. R. (2018). New insight into the early stages of biofilm formation. *Proceedings of the National Academy of Sciences*, 4317-4319.

Atshan, S. S., Nor Shamsudin, M., Sekawi, Z., Lung, L. T. T., Hamat, R. A., Karunanidhi, A., Mateg Ali, A., Ghaznavi-Rad, E., Ghasemzadeh-Moghaddam, H. and Chong Seng, J. S. (2012). Prevalence of adhesion and regulation of biofilm-related genes in different clones of *Staphylococcus aureus*. *Journal of BioMedicine and biotechnology*.

Badapanda, C. and Chikara, S. K. (2016). Lepidopteran antimicrobial peptides (AMPs): overview, regulation, modes of action, and therapeutic potentials of insect-derived AMPs. *Short Views on Insect Genomics and Proteomics*, 141-163.

Bahar, A. A. and Ren, D. (2013). Antimicrobial peptides. *Pharmaceuticals*, 1543-1575.

Bals, R. (2000). Epithelial antimicrobial peptides in host defense against infection. *Respiratory research*, 1-10.

Band, V. I. and Weiss, D. S. (2015). Mechanisms of antimicrobial peptide resistance in Gram-negative bacteria. *Antibiotics*, 18-41.

Bartlett, T. C., Cuthbertson, B. J., Shepard, E. F., Chapman, R. W., Gross, P. S. and Warr, G. W. (2002). Crustins, homologues of an 11.5-kDa antibacterial peptide, from two species of penaeid shrimp, *Litopenaeus vannamei* and *Litopenaeus setiferus*. *Marine Biotechnology*, 278-293.

Batoni, G., Maisetta, G., Lisa Brancatisano, F., Esin, S. and Campa, M. (2011). Use of antimicrobial peptides against microbial biofilms: advantages and limits. *Current Medicinal Chemistry*, 256-279.

Bechinger, B. and Lohner, K. (2006). Detergent-like actions of linear amphipathic cationic antimicrobial peptides. *Biochimica et Biophysica Acta (BBA)-Biomembranes*, 1529-1539.

Bílikova, K., Huang, S. C., Lin, I. P., Šimuth, J. and Peng, C. C. (2015). Structure and antimicrobial activity relationship of royalisin, an antimicrobial peptide from royal jelly of *Apis mellifera*. *Peptides*, 190-196.

Bjarnsholt, T. (2013). The role of bacterial biofilms in chronic infections. *APMIS*, 1-58.

Bjarnsholt, T., Jensen, P. Ø., Moser, C. and Høiby, N. (2011). *Biofilm infections*. Springer.

Brogden, K. A. (2005). Antimicrobial peptides: pore formers or metabolic inhibitors in bacteria? *Nature Reviews Microbiology*, 238-250.

Bulet, P. and Stocklin, R. (2005). Insect antimicrobial peptides: structures, properties and gene regulation. *Protein and peptide letters*, 3-11.

Buonocore, F., Fausto, A. M., Pelle, G. D., Roncevic, T., Gerdol, M. and Picchietti, S. (2021). Attacins: A Promising Class of Insect Antimicrobial Peptides. *Antibiotics*, 212.

Bylund, J., Christophe, T., Boulay, F., Nyström, T., Karlsson, A. and Dahlgren, C. (2001). Proinflammatory activity of a cecropin-like antibacterial peptide from *Helicobacter pylori*. *Antimicrobial agents and chemotherapy*, 1700-1704.

Čeřovský, V., Žďárek, J., Fučík, V., Monincová, L., Voburka, Z. and Bém, R. (2010). Lucifensin, the long-sought antimicrobial factor of medicinal maggots of the blowfly *Lucilia sericata*. *Cellular and Molecular Life Sciences*, 455-466.

Chen, X., Zhang, L., Wu, Y., Wang, L., Ma, C., Xi, X., Bininda-Emonds, O. R., Shaw, C., Chen, T. and Zhou, M. (2018). Evaluation of the bioactivity of a mastoparan peptide from wasp venom and of its analogues designed through targeted engineering. *International journal of biological sciences*, 599.

Ciofu, O., Tolker-Nielsen, T., Jensen, P. Ø., Wang, H. and Høiby, N. (2015). Antimicrobial resistance, respiratory tract infections and role of biofilms in lung infections in cystic fibrosis patients. *Advanced Drug Delivery Reviews*, 7-23.

Da Silva, P., Jouvensal, L., Lamberty, M., Bulet, P., Caille, A. and Vovelle, F. (2003). Solution structure of termicin, an antimicrobial peptide from the termite *Pseudacanthotermes spiniger*. *Protein Science*, 438-446.

Danihlík, J., Aronstein, K. and Petřivalský, M. (2015). Antimicrobial peptides: a key component of honey bee innate immunity: Physiology, biochemistry, and chemical ecology. *Journal of Apicultural Research*, 123-136.

Das, T., Sharma, P. K., Busscher, H. J., Van Der Mei, H. C. and Krom, B. P. (2010). Role of extracellular DNA in initial bacterial adhesion and surface aggregation. *Applied and environmental microbiology*, 3405-3408.

Davies, D. (2003). Understanding biofilm resistance to antibacterial agents. *Nature reviews Drug discovery*, 114-122.

De La Fuente-Núñez, C., Korolik, V., Bains, M., Nguyen, U., Breidenstein, E. B., Horsman, S., Lewenza, S., Burrows, L. and Hancock, R. E. (2012). Inhibition of bacterial biofilm formation and swarming motility by a small synthetic cationic peptide. *Antimicrobial agents and chemotherapy*, 2696-2704.

De la Fuente-Núñez, C., Reffuveille, F., Haney, E. F., Straus, S. K. and Hancock, R. E. (2014). Broad-spectrum anti-biofilm peptide that targets a cellular stress response. *PLoS pathogens*, e1004152.

de la Torre, B. G. and Albericio, F. (2020). *Peptide Therapeutics* 2.0. In (2293): Multidisciplinary Digital Publishing Institute.

Dean, S. N., Bishop, B. M. and Van Hoek, M. L. (2011). Natural and synthetic cathelicidin peptides with anti-microbial and anti-biofilm activity against Staphylococcus aureus. *BMC microbiology*, 1-13.

Di Somma, A., Moretta, A., Canè, C., Cirillo, A. and Duilio, A. (2020). Inhibition of bacterial biofilm formation. In *Bacterial Biofilms*. IntechOpen.

Donlan, R. M. (2001). Biofilm formation: a clinically relevant microbiological process. *Clinical infectious diseases*, 1387-1392.

Dosler, S. and Karaaslan, E. (2014). Inhibition and destruction of Pseudomonas aeruginosa biofilms by antibiotics and antimicrobial peptides. *Peptides*, 32-37.

Fehlbaum, P., Bulet, P., Michaut, L., Lagueux, M., Broekaert, W. F., Hetru, C. and Hoffmann, J. A. (1994). Insect immunity. Septic injury of Drosophila induces the synthesis of a potent antifungal peptide with sequence homology to plant antifungal peptides. *Journal of Biological Chemistry*, 33159-33163.

Fenner, A. (2020). Antimicrobial peptide derived from moths can eradicate UPEC biofilms and could offer a novel therapeutic option. *Nature Reviews Urology*, 191-191.

Flores-Mireles, A. L., Walker, J. N., Caparon, M. and Hultgren, S. J. (2015). Urinary tract infections: epidemiology, mechanisms of infection and treatment options. *Nature reviews microbiology*, 269-284.

Forman, M. E., Jennings, M. C., Wuest, W. M. and Minbiole, K. P. (2016). Building a better quaternary ammonium compound (QAC): branched tetracationic antiseptic amphiphiles. *ChemMedChem*.

Garrett, T. R., Bhakoo, M. and Zhang, Z. (2008). Bacterial adhesion and biofilms on surfaces. *Progress in Natural Science*, 1049-1056.

Gomes, B., Augusto, M. T., Felício, M. R., Hollmann, A., Franco, O. L., Gonçalves, S. and Santos, N. C. (2018). Designing improved active peptides for therapeutic approaches against infectious diseases. *Biotechnology advances*, 415-429.

Hara, S. and Yamakawa, M. (1995). Moricin, a Novel Type of Antibacterial Peptide Isolated from the Silkworm, *Bombyx mori* (∗). *Journal of Biological Chemistry*, 29923-29927.

Hara, S. and Yamakawa, M. (1996). Production in Escherichia coli of moricin, a novel type antibacterial peptide from the silkworm, *Bombyx mori*. *Biochemical and biophysical research communications*, 664-669.

Hemmi, H., Ishibashi, J., Hara, S. and Yamakawa, M. (2002). Solution structure of moricin, an antibacterial peptide, isolated from the silkworm *Bombyx mori*. *FEBS letters*, 33-38.

Herbert, S., Bera, A., Nerz, C., Kraus, D., Peschel, A., Goerke, C., Meehl, M., Cheung, A. and Götz, F. (2007). Molecular basis of resistance to muramidase and cationic antimicrobial peptide activity of lysozyme in staphylococci. *PLoS pathogens*, e102.

Høiby, N. (2017). A short history of microbial biofilms and biofilm infections. *Apmis*, 272-275.

Hollmann, A., Martinez, M., Maturana, P., Semorile, L. C. and Maffia, P. C. (2018). Antimicrobial peptides: interaction with model and biological membranes and synergism with chemical antibiotics. *Frontiers in chemistry*, 204.

Hwang, I. S., Hwang, J. S., Hwang, J. H., Choi, H., Lee, E., Kim, Y. and Lee, D. G. (2013). Synergistic effect and antibiofilm activity between the antimicrobial peptide coprisin and conventional antibiotics against opportunistic bacteria. *Current microbiology*, 56-60.

Hwang, J. S., Lee, J., Kim, Y. J., Bang, H. S., Yun, E. Y., Kim, S. R., Suh, H. J., Kang, B. R., Nam, S. H. and Jeon, J. P. (2009). Isolation and characterization of a defensin-like peptide (coprisin) from the dung beetle, *Copris tripartitus*. *International journal of peptides*.

Imler, J. L. and Bulet, P. (2005). Antimicrobial peptides in Drosophila: structures, activities and gene regulation. *Mechanisms of epithelial defense*, 1-21.

Jamal, M., Ahmad, W., Andleeb, S., Jalil, F., Imran, M., Nawaz, M. A., Hussain, T., Ali, M., Rafiq, M. and Kamil, M. A. (2018). Bacterial biofilm and associated infections. *Journal of the Chinese Medical Association*, 7-11.

James, G. A., Swogger, E., Wolcott, R., Pulcini, E. d., Secor, P., Sestrich, J., Costerton, J. W. and Stewart, P. S. (2008). Biofilms in chronic wounds. *Wound repair and regeneration*, 37-44.

Jenssen, H., Hamill, P. and Hancock, R. E. (2006). Peptide antimicrobial agents. *Clinical microbiology reviews*, 491-511.

Kalsy, M., Tonk, M., Hardt, M., Dobrindt, U., Zdybicka-Barabas, A., Cytrynska, M., Vilcinskas, A. and Mukherjee, K. (2020). The insect antimicrobial peptide cecropin A disrupts uropathogenic *Escherichia coli* biofilms. *NPJ biofilms and microbiomes*, 1-8.

Kaneko, Y., Furukawa, S., Tanaka, H. and Yamakawa, M. (2007). Expression of antimicrobial peptide genes encoding Enbocin and Gloverin isoforms in the silkworm, *Bombyx mori*. *Bioscience, Biotechnology, and Biochemistry*, 0708030532-0708030532.

Kim, S. R., Hong, M. Y., Park, S. W., Choi, K. H., Yun, E. Y., Goo, T. W., Kang, S. W., Suh, H. J., Kim, I. and Hwang, J. S. (2010). Characterization and cDNA cloning of a cecropin-like antimicrobial peptide, papiliocin, from the swallowtail butterfly, *Papilio xuthus*. *Molecules and cells*, 419-423.

Koo, H., Allan, R. N., Howlin, R. P., Stoodley, P. and Hall-Stoodley, L. (2017). Targeting microbial biofilms: current and prospective therapeutic strategies. *Nature reviews microbiology*, 740-755.

Koo, H. B. and Seo, J. (2019). Antimicrobial peptides under clinical investigation. *Peptide Science*, e24122.

Krizsan, A., Volke, D., Weinert, S., Sträter, N., Knappe, D. and Hoffmann, R. (2014). Insect-derived proline-rich antimicrobial peptides kill bacteria by inhibiting bacterial protein translation at the 70 S ribosome. *Angewandte Chemie International Edition*, 12236-12239.

Lee, M., Bang, K., Kwon, H. and Cho, S. (2013). Enhanced antibacterial activity of an attacin-coleoptericin hybrid protein fused with a helical linker. *Molecular biology reports*, 3953-3960.

Lehrer, R. I. and Ganz, T. (1999). Antimicrobial peptides in mammalian and insect host defence. *Current opinion in immunology*, 23-27.

Lele, D. S., Talat, S., Kumari, S., Srivastava, N. and Kaur, K. J. (2015). Understanding the importance of glycosylated threonine and stereospecific action of Drosocin, a Proline rich antimicrobial peptide. *European journal of medicinal chemistry*, 637-647.

Liaqat, I., Bachmann, R. T. and Edyvean, R. G. (2014). Type 2 quorum sensing monitoring, inhibition and biofilm formation in marine microrganisms. *Current microbiology*, 342-351.

Liaqat, I., Gulab, B., Hanif, U., Sultan, A., Sadiqa, A., Zafar, U., Afzaal, M., Naseem, S., Akram, S. and Saleem, G. (2022). Honey Potential as Antibiofilm, Antiquorum Sensing and Dispersal Agent against Multispecies Bacterial Biofilm. *Journal of Oleo Science*, ess21199.

Magiorakos, A., Burns, K., Baño, J. R., Borg, M., Daikos, G., Dumpis, U., Lucet, J., Moro, M., Tacconelli, E. and Simonsen, G. S. (2017). Infection prevention and control measures and tools for the prevention of entry of carbapenem-resistant Enterobacteriaceae into healthcare settings: guidance from the European Centre for Disease Prevention and Control. *Antimicrobial Resistance & Infection Control*, 1-17.

Manabe, T. and Kawasaki, K. (2017). D-form KLKLLLLLKLK-NH 2 peptide exerts higher antimicrobial properties than its L-form counterpart via an association with bacterial cell wall components. *Scientific Reports*, 1-10.

Manniello, M., Moretta, A., Salvia, R., Scieuzo, C., Lucchetti, D., Vogel, H., Sgambato, A. and Falabella, P. (2021). Insect antimicrobial peptides: potential weapons to counteract the antibiotic resistance. *Cellular and Molecular Life Sciences*, 1-24.

Matsuzaki, K. (2019). Membrane permeabilization mechanisms. *Antimicrobial Peptides*, 9-16.

Memariani, H., Memariani, M. and Pourmand, M. R. (2018). Venom-derived peptide Mastoparan-1 eradicates planktonic and biofilm-embedded methicillin-resistant *Staphylococcus aureus* isolates. *Microbial pathogenesis*, 72-80.

Memariani, H., Memariani, M., Shahidi-Dadras, M., Nasiri, S., Akhavan, M. M. and Moravvej, H. (2019). Melittin: from honeybees to superbugs. *Applied microbiology and biotechnology*, 3265-3276.

Mukhopadhyay, S., Prasad, A. B., Mehta, C. H. and Nayak, U. Y. (2020). Antimicrobial peptide polymers: No escape to ESKAPE pathogens—A review. *World Journal of Microbiology and Biotechnology*, 1-14.

Munita, J. and Arias, C. (2016). Virulence mechanisms of bacterial pathogens. *Microbiology Spectrum*, 481-511.

Mura, M., Wang, J., Zhou, Y., Pinna, M., Zvelindovsky, A. V., Dennison, S. R. and Phoenix, D. A. (2016). The effect of amidation on the behaviour of antimicrobial peptides. *European Biophysics Journal*, 195-207.

Neiger, M. R., González, J. F., Gonzalez-Escobedo, G., Kuck, H., White, P. and Gunn, J. S. (2019). Pathoadaptive alteration of Salmonella biofilm formation in response to the gallbladder environment. *Journal of bacteriology*, e00774-00718.

Otvos, J., Laszlo. (2000). Antibacterial peptides isolated from insects. *Journal of peptide science: an official publication of the European Peptide Society*, 497-511.

Overhage, J., Campisano, A., Bains, M., Torfs, E. C., Rehm, B. H. and Hancock, R. E. (2008). Human host defense peptide LL-37 prevents bacterial biofilm formation. *Infection and immunity*, 4176-4182.

Paczosa, M. K. and Mecsas, J. (2016). Klebsiella pneumoniae: going on the offense with a strong defense. *Microbiology and Molecular Biology Reviews*, 629-661.

Percival, S. L., Hill, K. E., Williams, D. W., Hooper, S. J., Thomas, D. W. and Costerton, J. W. (2012). A review of the scientific evidence for biofilms in wounds. *Wound repair and regeneration*, 647-657.

Piperaki, E. T., Syrogiannopoulos, G. A., Tzouvelekis, L. S. and Daikos, G. L. (2017). *Klebsiella pneumoniae*: virulence, biofilm and antimicrobial resistance. *The Pediatric infectious disease journal*, 1002-1005.

Reddy, K., Yedery, R. and Aranha, C. (2004). Antimicrobial peptides: premises and promises. *International Journal of Antimicrobial Agents*, 536-547.

Romoli, O., Mukherjee, S., Mohid, S. A., Dutta, A., Montali, A., Franzolin, E., Brady, D., Zito, F., Bergantino, E. and Rampazzo, C. (2019). Enhanced silkworm cecropin B antimicrobial activity against *Pseudomonas aeruginosa* from single amino acid variation. *ACS infectious diseases*, 1200-1213.

Rossi, E., Cimdins, A., Lüthje, P., Brauner, A., Sjöling, Å., Landini, P. and Römling, U. (2018). "It'sa gut feeling"–Escherichia coli biofilm formation in the gastrointestinal tract environment. *Critical reviews in microbiology*, 1-30.

Sahoo, A., Swain, S. S., Behera, A., Sahoo, G., Mahapatra, P. K. and Panda, S. K. (2021). Antimicrobial Peptides derived from Insects offer a novel therapeutic option to combat biofilm: A review. *Frontiers in Microbiology*.

Schlievert, P. M. and Peterson, M. L. (2012). Glycerol monolaurate antibacterial activity in broth and biofilm cultures. *Plos one*, e40350.

Schuhmann, B., Seitz, V., Vilcinskas, A. and Podsiadlowski, L. (2003). Cloning and expression of gallerimycin, an antifungal peptide expressed in immune response of greater wax moth larvae, Galleria mellonella. *Archives of Insect Biochemistry and Physiology: Published in Collaboration with the Entomological Society of America*, 125-133.

Scocchi, M., Tossi, A. and Gennaro, R. (2011). Proline-rich antimicrobial peptides: converging to a non-lytic mechanism of action. *Cellular and Molecular Life Sciences*, 2317-2330.

Seufi, A. M., Hafez, E. E. and Galal, F. H. (2011). Identification, phylogenetic analysis and expression profile of an anionic insect defensin gene, with antibacterial activity, from bacterial-challenged cotton leafworm, Spodoptera littoralis. *BMC Molecular Biology*, 1-14.

Shrivastava, S. R., Shrivastava, P. S. and Ramasamy, J. (2018). World health organization releases global priority list of antibiotic-resistant bacteria to guide research, discovery, and development of new antibiotics. *Journal of Medical Society*, 76.

Sobel, J. D. (2000). Bacterial vaginosis. *Annual review of medicine*, 349-356.

Sojka, M., Valachova, I., Bucekova, M. and Majtan, J. (2016). Antibiofilm efficacy of honey and bee-derived defensin-1 on multispecies wound biofilm. *Journal of medical microbiology*, 337-344.

Stoodley, P., Sauer, K., Davies, D. G. and Costerton, J. W. (2002). Biofilms as complex differentiated communities. *Annual Reviews in Microbiology*, 187-209.

Teerapo, K., Roytrakul, S., Sistayanarain, A. and Kunthalert, D. (2019). A scorpion venom peptide derivative BmKn–22 with potent antibiofilm activity against *Pseudomonas aeruginosa*. *PloS One*, e0218479.

Tenke, P., Kovacs, B., Jäckel, M. and Nagy, E. (2006). The role of biofilm infection in urology. *World journal of urology*, 13-20.

Tiwari, S. K., Sutyak Noll, K., Cavera, V. L. and Chikindas, M. L. (2015). Improved antimicrobial activities of synthetic-hybrid bacteriocins designed from enterocin E50-52 and pediocin PA-1. *Applied and Environmental Microbiology*, 1661-1667.

Toval, F., Köhler, C. D., Vogel, U., Wagenlehner, F., Mellmann, A., Fruth, A., Schmidt, M. A., Karch, H., Bielaszewska, M. and Dobrindt, U. (2014). Characterization of Escherichia coli isolates from hospital inpatients or outpatients with urinary tract infection. *Journal of clinical microbiology*, 407-418.

Ueda, K., Imamura, M., Saito, A. and Sato, R. (2005). Purification and cDNA cloning of an insect defensin from larvae of the longicorn beetle, *Acalolepta luxuriosa. Applied entomology and zoology*, 335-345.

Vestby, L. K., Grønseth, T., Simm, R. and Nesse, L. L. (2020). Bacterial biofilm and its role in the pathogenesis of disease. *Antibiotics*, 59.

Vizioli, J., Bulet, P., Charlet, M., Lowenberger, C., Blass, C., Müller, H. M., Dimopoulos, G., Hoffmann, J., Kafatos, F. and Richman, A. (2000). Cloning and analysis of a cecropin gene from the malaria vector mosquito, *Anopheles gambiae. Insect molecular biology*, 75-84.

Volkoff, A. N., Rocher, J., d'Alençon, E., Bouton, M., Landais, I., Quesada-Moraga, E., Vey, A., Fournier, P., Mita, K. and Devauchelle, G. (2003). Characterization and transcriptional profiles of three Spodoptera frugiperda genes encoding cysteine-rich peptides. A new class of defensin-like genes from lepidopteran insects? *Gene*, 43-53.

Watnick, P. and Kolter, R. (2000). Biofilm, city of microbes. *Journal of bacteriology*, 2675-2679.

Wu, Q., Patočka, J. and Kuča, K. (2018). Insect antimicrobial peptides, a mini review. *Toxins*, 461.

Yi, H. Y., Chowdhury, M., Huang, Y. D. and Yu, X. Q. (2014). Insect antimicrobial peptides and their applications. *Applied microbiology and biotechnology*, 5807-5822.

Zasloff, M. (2019). Antimicrobial peptides of multicellular organisms: my perspective. *Antimicrobial Peptides*, 3-6.

Zhang, J., Li, W., Chen, J., Qi, W., Wang, F. and Zhou, Y. (2018). Impact of biofilm formation and detachment on the transmission of bacterial antibiotic resistance in drinking water distribution systems. *Chemosphere*, 368-380.

Zhang, L. J. and Gallo, R. L. (2016). Antimicrobial peptides. *Current Biology*, R14-R19.

Zhang, Z. T. and Zhu, S. Y. (2009). Drosomycin, an essential component of antifungal defence in Drosophila. *Insect molecular biology*, 549-556.

Zhou, X. X., Wang, Y. B., Pan, Y. J. and Li, W. F. (2008). Nisin-controlled extracellular production of apidaecin in *Lactococcus lactis*. *Applied Microbiology and Biotechnology*, 947-953.

Chapter 7

The Use of Nanotechnology as an Effective Antibiofilm Approach: A Review

Faiza Bashir[1] and Iram Liaqat[1*]
[1]Department of Zoology, Government College University, Lahore, Pakistan

Abstract

Nanotechnology is an emerging science and nano-materials with extremely smaller size, large surface area, and volume to mass ratio confer idiosyncratic properties. Nanocomposites and nanoparticles (NPs) can have peculiar antimicrobial and anti-biofilm characteristics and are superior candidates for abolishing biofilms. Their intricate organization is key to the virulent nature of biofilms in the infectious mode because it provides resilience to traditional biocides. Pathogenic microbes have toxic potential to trigger various ailments among living beings ever since the advent of mankind, and this can be devastating at times. Chronic illnesses caused by biofilms account for 80% of infections in humans. Roughly 60% of diseases related to health care centers are caused by medical devices as they aid the direct colonization of varied microbes in different body organs. To combat biofilms, antibiotic therapy is frequently inadequate as a consequence of the high mutation rate in microbes, limited adherence of antimicrobials, acidic nature, low oxygen availability, and elevated enzymatic activities in biofilms. The implementation of nano- improved lubricants could be a convenient approach to hinder biofilm formation. Nano constructed antimicrobials have recently been found to eradicate planktonic and antibiotic-impervious bacteria. Metallic NPs have also been reported to suppress communities of microbes. Antimicrobial drug distribution via nano-

*Corresponding Author's Email: iramliaq@hotmail.com.

In: Understanding Antibiofilm Activity
Editor: Iram Liaqat
ISBN: 978-1-68507-927-7

materials has efficiently responded against biofilms. Rapid innovations in the domains of bio-sciences as well as gradual advancement in the sphere of nanotechnology have enabled nanotechnologists to configure new nano-modified antibiofilm materials and interfaces with efficient capabilities to knock out biofilms. In this chapter, we will examine a range of recently discovered nano biomaterials and explain the methodologies that are being employed to generate antibiofilm surfaces.

Keywords: nanotechnology, nanomaterials, biofilm, antimicrobials, toxic potential, antibiofilm

Introduction

Bacterial biofilms are perfect congregations of adhesive bacterial cells that have bonded to an inanimate or active surface as well as enveloped by an external polysaccharide matrix (EPS). This matrix is primarily made up of free DNA, carbohydrates (polysaccharides), and proteins (Costerton, 1999). The bacterial communities that proliferate in a biofilm are not exclusive to those that thrive in a normal environment. Morphological modifications, biochemical pathways, resistance to antimicrobials, and other traits distinguish biofilms from their planktonic counterparts (Mirani et al., 2018; Bimanand et al., 2018; Wang et al., 2016; Fulaz et al., 2019). Biofilms can be made up of microbes from the same or distinct species, and they can develop on any organic or inorganic substrate that can provide nutrients (Kassen and Rainey, 2004). Biofilm genesis allows bacteria to survive harsh environments. Moreover, the biofilm shelters them from the immune system of the host as well as antibiotic therapy.

Biofilms are indeed a significant threat to all biota and are extremely difficult to eradicate. The Center for Disease Control and Prevention estimate that roughly 1.7 million hospital-acquired illnesses were induced by biofilm forming bacteria in 2007, which resulted in financial damage of 11 billion U.S. dollars (Römling et al., 2014). Furthermore, biofilm-linked disorders cause approximately 0.5 million fatalities each year, with healthcare expenditure estimated to be around 94 billion dollars (Wolcott et al., 2010). In patients with cystic fibrosis and skin infections, *Enterococcus, Pseudomonas aeruginosa,* and *Staphylococcus* illnesses are highly prevalent (Römling et al., 2014). Biofilm-organizing bacteria are approximately 1000 times more impervious to antibiotics in contrast to planktonic bacteria (Soleimani et al., 2015). Biofilms exhibit extraordinarily flexible structural features, and their gene expression constantly alters, leading to changes in their associated antigens (Thoendel et al., 2011). Because of the significant concentration of metallic ions and lessened pH within the biofilm, the EPS layer prevents most types of migrations, which eventually lead to antibiotic deactivation. All of these features contribute to the ability of bacterial cells to withstand antibiotics when in a biofilm (Soleimani et al., 2015).

Nanotechnology is emerging science and is predominantly the characterization, mass production, and application of materials classified in a nanometer hierarchy. Nanoparticles (NPs) have an extremely small size (0.1-100 nm), large surface area, and volume to mass ratio and confer idiosyncratic properties, *i.e.,* magnetic, chemical, optical, and mechanical characteristics. Nanomaterials are gaining popularity in a variety of fields due to their increased functionality. An even more appealing feature of these nano-sized materials is the ability to manipulate their Physico-chemical characteristics, which is now being used in the biomedical field. Nanoparticles are being applied to combat infectious biofilms as an efficient alternative to antibiotics. Metal polymeric materials, proteins, or a fusion of both, high-intensity nano-structures and antibiotic coated nano-materials can have antibacterial activity while causing negligible damage to the host body (Lee and Ramasamy, 2016; Tufail and Liaqat, 2021).

Drug-packed nanoparticles can be employed to circumvent the drawbacks of traditional therapies, such as cytotoxicity, poor circulation, and metabolic breakdown. Nano-biotechnology is a discipline that integrates technology and nano-structures for innovative modification of biofilms amongst other applications. Nanotechnology can be used for sensing pathogenic thresholds, either by directly detecting the development of a particular microbial agent or by indirectly recognizing microbes by gauging their bioactivity. For example by tracking the frequency at which nutrients are being devoured (Kaittanis et al., 2008). Artificial and natural nano-materials have exhibited remarkable antibacterial characteristics, TiO_2, Ag nanoparticles, and chitosan are just a few. Chitosan is generated from chitin that is a natural polymer found in large quantities in arthropods and has recently been synthesized into nanoparticles (Qi et al., 2004).

Due to their peculiar Physico-chemical characteristics, the effects of metallic nanoparticles such as silver, platinum, palladium, and gold are being studied. Gold nanoparticles are utilized in the manufacture of biomarkers, on account of their potential to freeze biomolecules (Du et al., 2007). Antimicrobials can be incorporated into nanoparticles via direct embedding, adhesion, or chemical ligation, which has multiple benefits over the free product, including massively improved antibacterial activity and prolonged regulated antimicrobial distribution (Zhang et al., 2010). The mechanisms behind the antibiofilm potential of varied nano-materials, their advantages and disadvantages will be covered in this chapter.

Strategies to Combat Biofilms

Bacterial survival in biofilms is becoming a severe hazard to public health. Because of their potential to create biofilms, these bacteria have increased their pathogenicity (Bjarnsholt, 2013). As a result, prevention of biofilm growth and development is critical. Avoiding bacterial attachment to the substratum, using biofilm disrupting chemicals, and degrading the produced biofilm at an early stage are some of the tactics currently being used against biofilms. Infections obtained through the use of medical devices are typically treated with antimicrobials. These antibiotics are based on conventional susceptibility test findings on planktonic cells, but they frequently fail to produce the required results on biofilms, necessitating surgical removal of the implants (Nadell et al., 2008; Liaqat et al., 2014).

Several approaches, such as maintaining good hygiene and adopting antibiotic prophylaxis, are commonly used to regulate and prevent biofilm formation on medical devices. Nonetheless, antibiotic therapy should be carried out with caution to avoid the growth of antibiotic-resistant microbes. Some novel materials resistant to microbial adhesion have been investigated to combat medical biofilm infections by changing the surface properties (Camargo et al., 2009). Surfaces have been chemically modified to hold antibiotics, including the incorporation of antimicrobial agents onto metals and surface polymers to produce highly smooth surfaces. However, Antibacterial medications must be given in large volumes on surfaces for this approach to succeed, secondary infections and toxicity are significant drawbacks and the vast majority of antibiotics have failed (Bayston et al., 2009).

The first step in microbial adhesion to a surface is the formation of an adsorbed protein layer. When PEG (polyethylene glycol) or a zwitterionic species is applied to a surface, the surface characteristics are altered, and this can result in antifouling surfaces (Dorling and Baker, 2013). The presence of polyethylene molecules with a high molecular weight and the addition of sulphonate units results in significant resistance to adhesion by *Escherichia coli, Staphylococcus epidermidis, P. aeruginosa, Candida tropicalis,* and

Candida albicans (Roosjen et al., 2004). Other surface modification techniques have been investigated. The majority of these entail interfering with quorum signaling. Antimicrobial peptides with therapeutic potential, nitric oxide, enzymes targeting the EPS, ultrasound, and the utilization of light-activated antimicrobial drugs are just a few current examples (Sousa et al., 2011).

Nanotechnology in Biofilm Control and Treatment

The antibacterial processes of nanoparticles are fundamentally distinct from typical antimicrobial agents, making them an appealing approach for biofilm management. Many studies have shown that some nanoparticles enable controlled and sustained medication release, as well as better stability and reduced drug toxicity (Sun et al., 2012). Drug-resistant biofilm infections can be avoided due to recent advances in nanotechnology. In other experiments, nanoparticle-coated surfaces have also been discovered to act as biofilm inhibitory agents (Taylor and Webster, 2011).

Nanoparticles penetrate microbial cell walls and the biofilm layer due to their small size, causing irreparable damage to cell membranes and DNA. Long plasma half-lives and a high surface-to-volume ratio make them potential candidates for effective drug loading and targeting (Suci et al., 2007; Tufail and Liaqat, 2021). Nanotechnology offers a variety of approaches for tackling the growing threat of biofilm infections.

Antimicrobial and Antibiofilm Mechanisms of Nanoparticles

The fundamental processes behind the antimicrobial activities of nanoparticles are not entirely understood, they range from the creation of oxidative stressors, free radicals to DNA disruptions. Particle size, morphology, surface characteristics, composition, cell membrane modifications, impairment of respiratory activity, lipid peroxidation, ROS production, DNA shredding, nitrosation of peptide thiols and metabolic abnormalities are all thought to play important roles in the antibacterial mechanisms of various nanoparticles (Hajipour et al., 2012). The following are some of the qualities of nanoparticles that have been identified as being responsible for antibiofilm capabilities (Tufail et al., 2022).

Size

Several studies show that the narrower the diameter, the higher the antibacterial action as nanoparticles of smaller sizes have a stronger ability to penetrate the bacterial cell membrane. These cause a reduction in cell viability, which prevents biofilm formation (Morones et al., 2005).

Shape

When various shapes of nanoparticles were investigated, energy-filtering transmission electron microscopy revealed significant modifications in the cell membrane, resulting in cell death. Compared to spherical and rod-shaped silver nanoparticles, triangular-shaped truncated silver nanoplates with a basal lattice plane showed a higher biocidal effect (Pal et al., 2007).

Surface Charge

The capacity of nanoparticles to attract negatively charged bacterial cell walls improves as their zeta potential rises. As a result, cell membrane rupture occurs, resulting in strong antibacterial activity (El Badawy et al., 2011).

Composition

The adhesion of bacteria to a substratum is one of the fundamental steps in biofilm formation. Surfaces with nano-roughness have been shown to reduce bacterial adhesions, and, as a result, biofilm formation (Lemire et al., 2013). These properties of nanoparticles are used to develop antibiofilm effects through a variety of mechanisms, including cell membrane alterations, respiratory activity impairment, membrane lipid peroxidation, ROS generation, DNA unwinding, nitrosation of protein thiols by nitric oxide-releasing nanoparticles, and metabolic pathway disruption (Beyth et al., 2015).

Nanotechnology-Based Biofilm Control and Effective Treatments

Nanotechnology-based techniques are thought to hold promise to eliminate drug-resistant biofilm illnesses caused by medical tools and biomaterials. The use of nanoparticle-coated surfaces to suppress biofilms has been described in a modest number of investigations (Fisher et al., 2015). Materials at the nano-scale have distinctive physico-chemical and biological capabilities, as well as behaviors like proportional fluctuations, which larger particles of the same material do not exhibit. Nanoparticles are sufficiently small to infiltrate microbial cell walls and biofilm strata, causing irreparable destruction in the plasma membranes and DNA. They also have extended plasma half-lives and high surface-to-volume rates, making it easier to load medicines and target organisms (Suci et al., 2007).

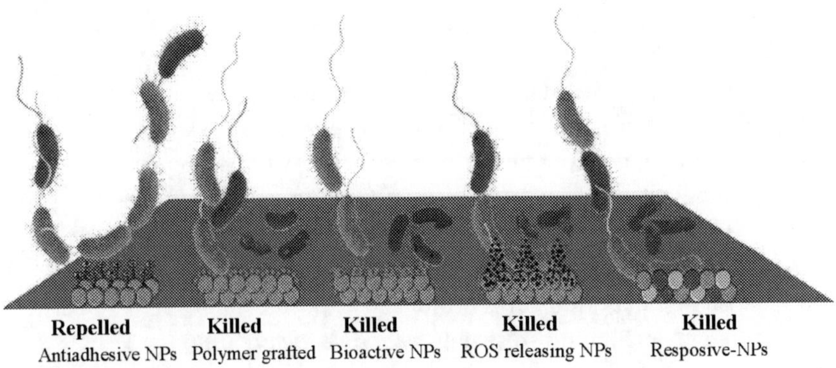

Figure 1. Effects of surface-engineered nanoparticles with distinct antimicrobial capabilities on biofilm inhibition.

Antibacterial Metals

Antibacterial and antibiofilm characteristics are found to be present in copper, gold, silver, titanium, and zinc, making them viable competitors over antibiotics without substantially raising the likelihood of the emergence of resistance. Impregnation strategies such as sintering, sandblasting, electro-deposition, spray coating, and electron beam evaporation improve the surface morphologies of metal-plated biomaterials. Consequently, microbial unions, protein immersions, and tissue integration features of these tools have been significantly modified making them more useful (Puckett et al., 2010).

Inorganic Nanoparticles

Different nanoparticles exhibit antibiofilm activity. Because of their remarkable antibacterial properties, silver nanoparticles (Ag NPs) have gained notable attention. Silver, whether ionic or as nanoparticles, has extraordinarily strong antimicrobial potential, making it extremely efficacious against pathogens linked with healthcare illnesses. Silver nanoparticles have a peculiar antibacterial strategy that involves a complicated interplay between silver ions and bacterial cell wall sulfhydryl groups, which disrupt bacterial cell membranes, enzyme activity, respiratory cycles, and cell differentiation (Sinha et al., 2011). By weakening intermolecular interactions, Ag NPs have been shown to break through biofilm matrices.

Recently, it has been revealed that biofilm production by *P. aeruginosa* and *S. epidermidis* was suppressed by more than 95% after 24 hours of treatment with Ag nanoparticles (Ansari et al., 2015). Silver-saturated hydroxyl-apatite and silver-titania composites have been reported to lessen bacterial colonization. They can obstruct Gram-positive and Gram-negative bacteria from establishing biofilms, while TiO_2 serves to prohibit silver agglomeration and permit the regulated emission of silver ions (Naik and Kowshik, 2014). Unfortunately, prolonged exposure to silver nanoparticles can encourage methicillin-resistant *Staphylococcus aureus* (MRSA) to establish silver tolerance, and large dosages of silver nanoparticles might cause wound repair to be delayed due to toxic impacts on skin cells (Kostenko et al., 2010).

Various metallic nanoparticles including zinc oxide (ZnO), copper oxide (CuO), titanium dioxide (TiO_2), iron oxide (Fe_2O_3), cerium oxide (CeO), magnesium oxide (MgO), and aluminum oxide (Al_2O_3) have all been examined for their antibacterial properties. In mammalian cells, ZnO nanoparticles have been demonstrated to have excellent antibacterial properties, very low lethality rate, and more efficient at suppressing biofilms of *Enterococcus faecalis, S. aureus, S. epidermidis, Bacillis subtilis,* and *E. coli* when compared to other metal oxides nanoparticles (Jones et al., 2008; Lee et al., 2014). In experiments on rats, ZnO nanoparticles in partnership with chitin coverings successfully cured skin wound ailments and suppressed biofilm formation. Additionally, the bacteriolytic and antibacterial capabilities of nano-textured ZnO have been demonstrated as superior to titania nanoparticles. However, zinc tolerance has been found in *P. aeruginosa* and *Proteus spp.* (Geilich and Webster, 2013; Zhang et al., 2013).

Due to the harmless qualities when compared to other metallic oxides, TiO_2 nanoparticles are classified as nontoxic antibacterial agents. They are commonly referred to as photocatalysts and are employed in a variety of areas, for example, self-cleaning and anti-fogging applications. Photocatalytic biofilm suppression by TiO_2 nanoparticles has been the subject of numerous studies. Furthermore, they have demonstrated strong antifungal biofilm action on clinical instrument surfaces particularly *C. albicans* (Haghighi et al., 2013). The antimicrobial effectiveness of TiO_2 nanoparticles is due to the generation of reactive oxygen species (ROS) in microbial cells as well as the oxidative damage to enzymes and increased lipid peroxidation which impairs metabolic performance and eventually leads to cell apoptosis (Ohko et al., 2009). Mesoporous TiO_2 nanoparticles have also been found to promote the liberation of linked bioactive compounds, resulting in extended anti-biofilm efficacy (Naik et al., 2013).

CuO nanoparticles display sufficient antimicrobial potential against a variety of bacteria, however, they are weaker antibacterial candidates when compared to silver and zinc nanoparticles, necessitating larger concentrations to accomplish similar antimicrobial effects. Unfortunately, CuO nanoparticles could be hazardous to mammalian cell lines at these higher levels (Ren et al., 2009). Nevertheless, these molecules have been reported to act remarkably against biofilms when coupled with some other nanoparticles. Nanoparticles of CuO coated with ZnO demonstrated significant biofilm suppressing efficacy in a tooth model (Eshed et al., 2012).

Iron nanoparticles are extensively employed as MRI contrast negotiators. However, at 8 nm, they have been proven to be effective in preventing *S. epidermidis* infection on orthopedic implants. Interestingly, antibiotic-linked magnetic iron nanoparticles outperform unconjugated magnetic iron nanoparticles against *E. faecalis* in both planktonic and biofilm phases (Chifiriuc et al., 2013). Catheters impregnated with 5 nm fundamental iron nanoparticles, exhibit biofilm immunity towards *S. aureus* and *P. aeruginosa*, and they may be used *in vivo* (Anghel et al., 2012). Gold nanoparticles have minimal to no antimicrobial effect on their own. Despite this, when attached to antibiotics, cytotoxic chemicals, or bio-elements they exhibit significant bactericidal and biofilm suppressive characteristics against a wide range of microbes, particularly antibiotic and multi-drug resistant types (Mu et al., 2016). Gold nanoparticles are now being coupled with targeted compounds to provide selective antibiofilm functions since they are harmless to cells (Chen et al., 2010).

Organic Nanoparticles

Antibacterial capabilities are achieved by distributing antibiotics, antimicrobial chemicals, bacteriolytic proteins, or by altering the surface areas of polymeric nanoparticles and polymer-build implants with alkyl pyrimidines or quaternary ammonium composites to elicit proximal execution. Antimicrobial polycationic molecules induce cell degradation, presumably by an ion interchange contact between bacteria and charged polymer interfaces, leading to plasma membrane cracking (Lichter and Rubner, 2009).

E. coli antibiofilm effectiveness has been found for a nano-porous polymer formed of sodium dodecyl sulfate. Antibacterial and antifungal responses are also being observed in vitamin-E-linked cationic polymer conjoined bio-disposable hydrogels (Nevius et al., 2012). Efficient and enhanced antibiofilm behavior of Levofloxacin (antibiotic) conjugated poly(lactic-co-glycolic acid) nanoparticles encapsulated with phosphatidyl-choline nanocomposites manifest remarkable antibacterial activity against *E. coli* biofilms (Cheow et al., 2010). Intriguingly, a silicone synthesized poly-dimethyl-siloxane (PDMS) surface (labeled as a "brush layout") proved particularly efficient against *C. albicans, S. aureus,* and *E. coli* bacterial and fungal biofilms while not caused no mammalian toxicity (Li et al., 2014).

Additionally, by employing polymers, surfaces of titanium have been improved physico-chemically, for example polyethylene oxide, polymethacrylic acid, polyurethane acetate and polyethylene glycol have been coupled with titanium and modified in a similar way. They show efficient performance to hinder protein assimilation and bacterial adhesion in combating biofilms (Zhang et al., 2008). Nitric oxide producing silica nanoparticles have been used to treat biofilm-associated wound septicaemia in mice, where they induced remarkable reduction in bacterial burdens of methicillin-tolerant *S. aureus*, *A. baumannii* and *C. albicans* (Macherla et al., 2012).

Metal-Polymer Nanocomposites

Organic polymers have weak mechanical characteristics when used for medical devices. However, they can be deposited on metallic substrates by spin plating, dip covering, or layer-by-layer plasma deposition (Vasilev et al., 2009). Gallium and zinc nanoparticles integrated into a polyether urethane blended PEG framework have been found to suppress *P. aeruginosa* contamination in rats through the coordinated emission of gallium nanoparticles, although zinc nanoparticles on their own were less potent. Silver and various heavy metal nanoparticles have been shown to eradicate numerous Gram positive and Gram negative planktonic microbes. These nanoparticles are simple to make via physical, chemical, or biological processes. Treatments with a wide range of options are possible, for example by coating or emulsions. Moreover, they work synergistically with antibiotics. However, they are potentially cytotoxic (Ma et al., 2013).

Dendrimers

Dendrimers are structures with three facets that may enclose both polar and non-polar substances in the vacuum spaces of their highly branching architectures (Ramalingam et al., 2013). The antimicrobial activity of manufactured-low-molecular-weight protein dendrimers in the case of *E. coli* and *S. aureus* has been demonstrated in the absence of supplementary antibiotics (Janiszewska et al., 2003). According to other examinations, the binding of fucose-specific lectins (LecB) to fucosepeptide dendrimer ligands interrupts *P. aeruginosa* adhesion and prevents the production of biofilms (Johansson et al., 2008).

Cyclodextrins

Cyclodextrins (CDs) are glucopyranose-based cyclic chemical structures that are being utilized to liquefy hydrophobic substances in water. *C. albicans* biofilm generation has been 96 percent inhibited *in vitro* when CDs were surface organized with polyethylene or polypropylene (coated with miconazole). Importantly, fungicidal effectiveness against *C. albicans* biofilms has been demonstrated using gold surface activated CD mounted anidulafungin and thymol (Gharbi et al., 2012; Nava-Ortiz et al., 2010). When drug loading and retention were improved, ciprofloxacin-packed CD-agar nanogels had been found to exhibit significant antibacterial effectiveness against *E. coli, S. epidermidis, P. aeruginosa,* and *S. aureus* (Blanco-Fernandez et al., 2011).

Lipid-Based Nanoparticles and Microemulsions

As they bear a close resemblance to organic cell membranes, liposomes are being used in a variety of pharmacological procedures, notably biofilm-related medicines. During *in vitro* examinations on clinical strains of *Bordetella bronchiseptica, E. coli, Acinetobacter baumannii, P. aeruginosa, Klebsiella pneumoniae,* and *Acinetobacter lwoffii*, drug-stuffed liposomes exhibited potent biofilm downregulation and interrupted quorum monitoring along with lowered lipase and chitinase processing (Alhariri and Omri, 2013). Biofilm-organizing bacteria have been exterminated using solid lipid nanoparticle (SLN) compositions that comprise antimicrobial drugs. In a mouse model of *S. aureus* triggered mastitis, a solid lipid nanoparticle mixture carrying PVA saturated castor oil with tilmicosin was employed, and another SLN formulation with eugenol indicated antifungal efficacy in a rat model of oral candidiasis (Garg and Singh, 2011). By destabilizing cytoplasmic barriers, agglomerating cytoplasm, and shifting intrinsic metabolic processes, microemulsions have been revealed as dominant antibiofilm agents against *P. aeruginosa* and *C. albicans* (Al-Adham et al., 2003).

Carbon-Based Nanomaterials

The antimicrobial effects of graphene-based nanoparticles are largely determined by their size and form. Since these nanoparticles are hydrophobic with low water solubility and remarkable persistence, they are frequently used as surface contact-killing compounds (Zou et al., 2016). However, mounted on substratum surfaces, 1-dimensional nanostructures and 2-dimensional carbon nano-flakes primarily had modest contact-killing and limited biofilm inhibition. Because of the efficient electron transport characteristics, zero-dimensional graphene oxide quantum dots (GQDs) have more peroxidase-like performance than 2D-graphene. Peroxidase resembling GQDs have been inspected to generate h+ ions in the vicinity of lowered proportions of H_2O_2, prompting Gram-negative *E. coli* and Gram-positive *S. aureus* to lyse and preventing bacterial adhesion. Infected injury repair and oral biofilm prevention have both been suggested for graphene nano-material applications. When a mouse model with infected wounds was treated with GQD-packed wound-dressings, an improved antibacterial action combined with narrowed levels of H_2O_2 was found (Sun et al., 2014).

Carbon Quantum-Dots

Carbon quantum dots (CQDs) are also known as C-dots. They are nanoscale carbon molecules (usually 10 nm in diameter) that are typically manufactured using hydrothermal processes and can be manipulated with a variety of operational compounds. Twenty four CQDs have been manufactured from natural sources like polyamines and solid ammonia citrate, which exhibit excellent positive charge and antibacterial activity against planktonic Gram-positive *S. aureus* and *B. subtilis* and Gram-negative *E. coli* and *P. aeruginosa* microbes when supplemented with spermidine (Jian et al., 2020).

Carbon-QDs conjugated with spermidine induce bacterial cell membrane breakdown with minimal cytotoxicity and cell lysis. In a rabbit model having bacterial keratitis, spermidine-doped CQDs have been reported to effectively promote the breakdown of corneal epithelial tight junctions, thereby increasing the efficiency of topical therapy (Jian et al., 2017). Carbon quantum dots generated by burning *Lactobacillus plantarum* effectively hindered the creation of *E. coli* biofilms with minimal cell lethality. Amine-functionalized QCDs that have been made luminescent by utilizing lauryl betaine, exhibit bacterial scanning potential and can be employed to assess and pinpoint infection spots (Yang et al., 2016).

Responsive Smart Nanoparticles

Antimicrobial infections are now being controlled with a blend of exogenous energy and energy-engrossing nanoparticles. The primary strategy entails employing external sources of energy including visual light, heat, near-infrared radiation (NIR), or high intensity alternating magnetic fields to induce irreparable disruptions in diseased cells by upregulating metallic nanoparticles or polymer-associated systems (Wong et al., 2015; Zharov et al., 2006; Maity et al., 2010). Photothermal materials, for example gold, iron oxide, and graphene nanoparticles have been found to convert NIR light into temperature. Owing to their unique reactivity with NIR light sources, gold nanoparticles of diverse types have been investigated intensively, however, this sensitivity varies with particle size. When gold or graphene nanoparticles were subjected to NIR, Gram-positive, Gram-negative, and hybrid bacterial species were found substantially denatured. The temperatures in the nanoparticle treated bacterial cultures was found to be higher than the bacteria's normal physiological ranges (Zhu et al., 2014).

Photodynamic treatment (PDT) employing light and photosensitizers has also utilized nanosized vectors to eliminate microbes. When, photosensitizer-nanoparticle aggregates are exposed to light, cytotoxic reactive oxygen species are produced which elicit microbial disintegration in both planktonic and biofilm modes. The effectiveness of hybridizing photosensitizers on nanoparticles to eradicate pathogens and biofilms has been investigated. *C. albicans*, MRSA, and complex microbial biofilms have been demonstrated to be significantly blocked by nanoparticles complexed with porphyrin, methylene blue, and rose bengal (Khan et al., 2012; Shrestha and Kishen, 2014). Photodynamic therapy holds promise to cure septicaemia, yet it is currently not employed for clinical usage due to the need to determine physico-chemical parameters of photosensitizers, dose administration, light dosimetry, and drug release control.

Magnetic nanoparticles (MNPs) intercept electromagnetic radiations from high-intensity axial magnetic field (AMF) and effectively redistribute it as targeted heat, and this overheating has been used to destroy *S. aureus* and *P. aeruginosa* biofilms *in vitro* (Park et al., 2011). Magnetic nanoparticles heating has been reported to effectively prohibit *S. aureus* biofilms in both *in vitro* and *in vivo* cultures of a mouse model with epidermal wound infection (Kim et al., 2013).

Nanotechnology-Build Antimicrobial-Transporting Systems

Mesoporous Silica Nano-Carriers, Liposomes, Polymeric Nano-Carriers, and Dendrimeric Nano-Carriers

The antimicrobial characteristics of nanoparticles are not only inherent but can also be used as antimicrobial targeted delivery systems, especially when core-assembled. In this section, the benefits and drawbacks of various antimicrobial nano-carriers are reviewed.

Table 1. Antibacterial and antibiofilm capabilities of nanostructured materials

Substances	Nano-Material Interpretation	Antibiofilm Tools	Antimicrobial System of NPs	References
Inorganic Nanoparticles	Ag-NPs and Surface-synthesized Gold NPs (Au-NPs)	Urethral stents, Central venous catheters Ventricular drain stents	Released silver ion combines with bacteria's sulfhydryl groups, disrupting permeability of the cell membrane, enzymatic activity, metabolic chains, and cell differentiation. EPS system was disrupted by a strongly positive charge density	Kumar and Sujitha, (2014) Wu et al. (2015) Fisher et al. (2015) Giri et al. (2015)
Organic Nanoparticles	Quadruplicate ammonium-chitosan Nanoparticles PEG balanced Lipid Nanoparticles	Bone and oral cements/fillings	Larger charged polymer chains pass through the cell membrane and generate charge transformations, which impair biofilm genesis	Shi et al. (2006) Sun et al. (2012)
Metallic polymeric Nano-mixtures	Silver titanium (Ag-Ti) nano-fusions, Silver or antibiotic coated NPs, Silver coupled silicone NPs, Diamond-like carbon-metal nano-composites,	Face masks, Heart valve Catheter against fungi and Pedicle screws	The matrix of EPS gets damaged by elevated positive charge density. Ag ions bonded with DNA interfere with electron transfer systems, impair pathogens and disintegrate biofilms	Li et al. (2006) Zhang et al. (2013) Angelina et al. (2017) Grunkemeier et al. (2006) Andara et al. (2006)
Substances	Nano-Material Interpretation	Antibiofilm Tools	Antimicrobial System of NPs	References
	Silicone-antibiotic loaded liposome, Polymeric Ag-NPs, Ag-NPs coated surfaces, Poly-cationic nanoparticles			Walraven and Lee, (2013) Ho et al. (2013) Hazer et al. (2016) Atar-Froyman et al. (2015)
Metal-Polymer Nano-composites	ZnO-NP associated titanium implants,	Orthopedic implants	Protein adsorptions and intracellular processes are affected	Elizabeth et al. (2014)

	TiO_2 nanostructured deposits, Ag-NP bonded poly(ethylene glycol diacrylate)-co-acrylic acid (PEGDA-AA) nanogel coatings on titanium surface, Quaternary ammonium salts (QAS) loaded with TiO_2 nanobars, ciprofloxacin-burdened nano-chitosan coated titanium substrates, polymer-NP based photodynamic treatment		by ZnO. The negatively charged bacteria get shredded by the positive interface of QAS. Ciprofloxacin hinders enzymes like DNA gyrase when it is liberated. The bacterial breakdown is caused by topoisomerases. ROS or superoxide hydroxyl radicals are formed when free radicals react with endogenous molecular oxygen. While hydrogen peroxide disrupts the membrane permeability of bacteria, resulting in irreversible bacterium deterioration	Peng et al. (2013) X. Chen et al. (2013) De Giglio et al. (2012) De Freitas et al. (2016)
Metallic/metal-polymer-based Nano-compounds	Titanium coated surfaces with ZnO-NPs, ZnO coupled with antibiotics	Implanted teeth dental sealants and root canal system filling	Direct interaction with ZnO causes the production of reactive oxygen species (ROS). Gentamycin's irreversible binding disturbs microorganisms	Abdulkareem et al. (2015) Zhao et al. (2011) Sharma et al. (2016) Magalhães et al. (2012) Ahn et al. (2009)
Substances	Nano-Material Interpretation	Antibiofilm Tools	Antimicrobial System of NPs	References
Nano-metallic polymers	Fixatives silica nanogel with AgNPs composites, Zn coupled CuO NPs plating on contact lenses, Chitosan loaded AgNPs and rapheme oxide linked antifungal drug	Contact lenses	released Ag ions disintegrate the bacteria and inhibit biofilm development. Voriconazole inhibits ergosterol synthesis by inhibiting 14-alpha sterol demethylase which produce antifungal activity.	Qu et al. (2013) Bazzaz et al. (2014) Tuby et al. (2016) Huang et al. (2016)

Mesoporous silica nanoparticles are suitable candidates for transporting antimicrobials and other medicines. The electrodynamic dual-layer reactions between the antimicrobial drugs and silica layer modify the loading capabilities of antimicrobials to be delivered. Diversely loaded monomers can be employed to improve the compatibility of silica nanoparticles to transport alternatively charged antibiotics to obtain high drug encapsulation effectiveness. Mesoporous silica nanoparticles can hold up to 10 times more antibiotic than non-porous silica nanoparticles due to their large BET (Brunauer–Emmett–Teller), contact area, and permeability. However, antimicrobials transported in silica nanoparticles are not well restricted and can readily end up in the bloodstream on their journey to an infected location in the body.

Chemically coupled antibiotics are far less likely to be released accidentally. Proton-activated liberation of nitric oxide is achievable with N-diazeniumdiolates-based silica nanoparticles. Biofilm-cultured *S. epidermidis, P. aeruginosa, E. coli, S. aureus,* and *C. albicans* were destroyed by nitric oxide-producing silica nanoparticles, with only minor effects on fibroblast growth rates. Via free radical pathways comparable to the antimicrobial strategies of immune cytokines, nitric oxide not only eliminates but also disseminates biofilm occupants (Kwon et al., 2017). To maintain sufficient loading, water repellent antimicrobials like cinnamaldehyde or triclosan must be covalently bound to mesoporous silica nanoparticles. Cinnam-aldehyde attached to silica nanoparticles, when released under acidic circumstances, have excellent ability to remove infectious biofilms of methicillin-tolerant *S. aureus, P. aeruginosa, E. coli,* and *Enterobacter cloacae* while preferentially boosting fibroblast growth.

Triclosan-(3-(triethoxysilyl) propyl) carbamate-bound silica nanoparticles exhibit superior antibacterial activity in contrast to free triclosan against *E. coli* and *S. aureus* in a typical planktonic framework. Furthermore, last-resort hydrophilic antimicrobials such as vancomycin can be coupled with amine-operated silica nano-domains for simultaneous growth inhibition and translocation of bacteria (Duncan et al., 2015). Liposomes are organic mimics of polymersomes and bear a very striking resemblance in shape and form with hydrophilic centers, which make them suitable for delivering water-loving antibiotics, which micelles cannot transport. These nanostructures contain phospholipids and have the distinct ability to fuse with bacterial phospholipid bilayer, which causes their antimicrobial payload to be released directly into the bacterial cells. Most traditional antibiotics can be incorporated into liposomes because of their underlying hydrophilic center while combining basic and anionic liposomes with the phospholipid bilayers has shown enhanced mortality of many different bacterial strains in their biofilm phase of proliferation. Liposome integration does not result in membrane disintegration, but rather generates transporting routes due to their molecular homology with bacterial layers (Forier et al., 2014).

Pores are thought to allow passage through the peptidoglycan matrix. Antibiotic-stuffed phosphatidylcholine liposomes that have been cationically transformed are effectively captured by negatively charged bacterial plasma membranes, leading to greater infiltration into biofilms and efficient bacterial eradication as compared to neutral liposomes or passively dispersed antibiotics. Toxins or photocatalytic compounds have been integrated into liposomes to facilitate the controlled union with bacterial membranes and subsequent antimicrobial activity. Pore-generating toxicants produced by microorganisms can perhaps impair liposome walls and stimulate the liberation of packaged antimicrobial drugs. For example, a toxicant produced by methicillin-tolerant *S. aureus* will initiate the emission of vancomycin from Au-nanoparticle balanced liposomes to limit planktonic *S. aureus* development (Pornpattananangkul et al., 2011). The destruction caused by NIR irradiation on liposomes using photodynamic cypate also promotes cleavage and the discharge of enclosed antimicrobials (Zhao et al., 2018). Due to the vast variety of polymers available, polymeric nanoparticles present ample opportunities for modification to achieve improved targeting towards biofilms and responsive drug release.

Natural Polymeric Nanocarriers

Unorganized charged particles and antibacterial nanoparticles can be derived from organic polymers like chitosan or e-polylysine. Chitosan has been the most widely employed nanoparticle for this application. Chitosan is a polymeric compound made up of N-acetylglucosamine and glucosamine sequences that are ordered randomly. At pH 6.5, the majority of substituents on free glucosamine get protonated and are therefore attracted to the negative charge, as seen in pathogenic biofilm communities. By binding to DNA, positively charged chitosan units may engage with negatively charged bacterial components for example with superficial membranes of the bacteria, leading to membrane disruption, interior component efflux, suppression of cellular metabolism, and mRNA transmission. Chitosan exhibits a greater antibacterial potential against Gram-positive strains as compared to Gram-negative (Pelgrift and Friedman, 2013).

While intense perforation is restricted by electric bilayer contact occurring in a biofilm network, nanoparticles made of chitosan or its quaternary ammonia derivatives have a positively charged interface, which aids their deposition in biofilms. Chitosan nanocarriers can work in conjunction with their antibacterial payloads once they surpass barriers in a biofilm. Freight also contains photosensitizer substances like methylene blue, which when exposed to UV light produces harmful superoxide anionic oxygen. It also displays complementarity with chitosan NPs in the battle against *S. aureus* and *P. aeruginosa* biofilms. Recently, the antibacterial activity of vancomycin-loaded gelatin nanocarriers has been investigated in zebrafish larvae (Forier et al., 2014; Zhang et al., 2018).

Almost all of the obstacles to antibiotic administration inside a biofilm can be handled by purpose-built micelles, and prolonged travel in the bloodstream can also be achieved by customizing or choosing the water-favoring tail groups appropriately. The hydro-repellent center of micelles facilitates the loading of antimicrobial compounds that are weakly water dissolvable, such as ciprofloxacin, which attacks *S. aureus* and *P. aeruginosa* biofilms or carvacrol oil, which targets *P. aeruginosa* without harming fibroblast vitality in mixed-culture with bacteria (Landis et al., 2017). Surface-adjustable combined-shell polymeric micelles (MSPMs) made up of the copolymers show no consequential alternations in size or ionic capability after being treated with blood plasma and their zeta potential turns positive during infiltration through acidic biofilms, indicating the non-appearance of blood protein, allowing aggregation without washout (Liu et al., 2016).

In the acidic medium of *Streptococcus mutans* biofilms, farnesol-packed molecules disintegrate to liberate farnesol, displaying 4-times stronger cytotoxicity over farnesol in solution (Horev et al., 2015). Antimicrobials packaged in micellar nanostructures can be liberated by employing not just pH, but bacterial enzymes as well. Surface stabilized antimicrobial-burdened PEG-b-PCL and PCL-b-PAE MSPMs allow widespread infiltration and intensification of the MSPMs in staphylococcal biofilms, as well as electrostatic dual-layer action to attack bacterial substrates, where bacterial lipases debase the poly(e-caprolactone) circuits to withdraw the antimicrobial shipment. When compared to triclosan in solution and triclosan-packed in single shell polymeric micelles, triclosan-laden MSPMs have greater staphylococci combating capabilities in both their planktonic and biofilm phases of proliferation. The basic disadvantage of incorporating antimicrobials into micellar centers is that antimicrobial drugs may be lost in blood circulation on approach to the target spot. To overcome this drawback, antimicrobials can be conjugated into surfactant-forming micelles, through links that can be dissolved by bacterial enzymes once inside a biofilm (Liu et al., 2016; Liu et al., 2018).

Dendrimers have polar hydroxyl or amino groups that provide effective options for antibiotic integration via ester and amide bonds. Amino groups on dendrimers can generate amide bonds with carboxyl groups in antibiotics like vancomycin. When compared to vancomycin in solution, vancomycin-incorporated dendrimers show efficient binding capacity against vancomycin-tolerant *S. aureus*, because of the high density of vancomycin on the dendrimer surface. Similarly, PAMAM dendrimers that release nitric oxide greatly hinder *P. aeruginosa* and *S. aureus* biofilms (Carpenter and Schoenfisch, 2012).

Conclusion

Although there have been breakthroughs in the discovery of unique antibiofilm substances, existing biofilm therapeutic interventions are constrained by high prices and intricacies of use, thus, necessitating the exploration of cost-effective substitutes. As this review demonstrates, significant advancements in nanotechnology-based techniques for controlling, reducing, and eradicating bacterial biofilm illnesses, primarily of biomedical implants, have been made. For wound remedies and related clinical challenges, various nanoparticles and nanocomposites with established bactericidal and fungicidal capabilities have already been proven as suitable replacements to antibiotics. Coatings, biological toxins, drug-transporting systems, and implants, all can contain nano-substances, and development in these sectors is still ongoing. To guarantee successful therapeutic relevance however, fundamental concerns such as NP susceptibility and the surface interplay between NPs, biofilms, and the host must be addressed. When compared to ordinary antimicrobials, which have a narrow range of effectiveness and are ecologically hazardous, nanomaterials antibiofilm coatings on devices are likely to have persistent antimicrobial properties while being much less invasive.

References

Abdulkareem, E. H., Memarzadeh, K., Allaker, R., Huang, J., Pratten, J. and Spratt, D. (2015). Anti-biofilm activity of zinc oxide and hydroxyapatite nanoparticles as dental implant coating materials. *Journal of dentistry*, 1462-1469.

Ahn, S. J., Lee, S. J., Kook, J. K. and Lim, B. S. (2009). Experimental antimicrobial orthodontic adhesives using nanofillers and silver nanoparticles. *Dental Materials*, 206-213.

Al-Adham, I., Al-Hmoud, N., Khalil, E., Kierans, M. and Collier, P. J. (2003). Microemulsions are highly effective anti-biofilm agents. *Letters in Applied Microbiology*, 97-100.

Alhariri, M. and Omri, A. (2013). Efficacy of liposomal bismuth-ethanedithiol-loaded tobramycin after intratracheal administration in rats with pulmonary *Pseudomonas aeruginosa* infection. *Antimicrobial agents and chemotherapy*, 569-578.

Andara, M., Agarwal, A., Scholvin, D., Gerhardt, R. A., Doraiswamy, A., Jin, C., Narayan, R. J., Shih, C. C., Shih, C. M. and Lin, S. J. (2006). Hemocompatibility of diamondlike carbon–metal composite thin films. *Diamond and related materials*, 1941-1948.

Angelina, J. T. T., Ganesan, S., Panicker, T., Narayani, R., Paul Korath, M. and Jagadeesan, K. (2017). Pulsed laser deposition of silver nanoparticles on prosthetic heart valve material to prevent bacterial infection. *Materials Technology*, 148-155.

Anghel, I., Limban, C., Grumezescu, A. M., Anghel, A. G., Bleotu, C. and Chifiriuc, M. C. (2012). *In vitro* evaluation of anti-pathogenic surface coating nanofluid, obtained by combining Fe_3O_4/C_{12} nanostructures and 2-((4-ethylphenoxy) methyl)-N-(substituted-phenylcarbamothioyl)-benzamides. *Nanoscale research letters*, 1-10.

Ansari, M. A., Khan, H. M., Khan, A., Cameotra, S. and Alzohairy, M. (2015). Anti-biofilm efficacy of silver nanoparticles against MRSA and MRSE isolated from wounds in a tertiary care hospital. *Indian journal of medical microbiology*, 101-109.

Atar-Froyman, L., Sharon, A., Weiss, E. I., Houri-Haddad, Y., Kesler-Shvero, D., Domb, A. J., Pilo, R. and Beyth, N. (2015). Anti-biofilm properties of wound dressing incorporating nonrelease polycationic antimicrobials. *Biomaterials*, 141-148.

Bayston, R., Fisher, L. E. and Weber, K. (2009). An antimicrobial modified silicone peritoneal catheter with activity against both Gram positive and Gram negative bacteria. *Biomaterials*, 3167-3173.

Bazzaz, B. S. F., Khameneh, B., Jalili-Behabadi, M. M., Malaekeh-Nikouei, B. and Mohajeri, S. A. (2014). Preparation, characterization and antimicrobial study of a hydrogel (soft contact lens) material impregnated with silver nanoparticles. *Contact Lens and Anterior Eye*, 149-152.

Beyth, N., Houri-Haddad, Y. and Domb, A. (2015). Alternative antimicrobial approach: nanoantimicrobial materials. Evid-Based Compl Alt 2015: 1–16. In.

Bimanand, L., Taherikalani, M., Jalilian, F. A., Sadeghifard, N., Ghafourian, S., Mahdavi, Z., Mohamadi, S., Sayehmiri, K., Hematian, A. and Pakzad, I. (2018). Association between biofilm production, adhesion genes and drugs resistance in different SCCmec types of methicillin resistant *Staphylococcus aureus* strains isolated from several major hospitals of Iran. *Iranian journal of basic medical sciences*, 400.

Bjarnsholt, T. (2013). The role of bacterial biofilms in chronic infections. *Apmis*, 1-58.

Blanco-Fernandez, B., Lopez-Viota, M., Concheiro, A. and Alvarez-Lorenzo, C. (2011). Synergistic performance of cyclodextrin–agar hydrogels for ciprofloxacin delivery and antimicrobial effect. *Carbohydrate polymers*, 765-774.

Camargo, L., Marra, A., Büchele, G., Sogayar, A., Cal, R., de Sousa, J., Silva, E., Knobel, E. and Edmond, M. (2009). Double-lumen central venous catheters impregnated with chlorhexidine and silver sulfadiazine to prevent catheter colonisation in the intensive care unit setting: a prospective randomised study. *Journal of Hospital Infection*, 227-233.

Carpenter, A. W. and Schoenfisch, M. H. (2012). Nitric oxide release: Part II. Therapeutic applications. *Chemical Society Reviews*, 3742-3752.

Chen, W. Y., Lin, J. Y., Chen, W. J., Luo, L., Wei-Guang Diau, E. and Chen, Y. C. (2010). Functional gold nanoclusters as antimicrobial agents for antibiotic-resistant bacteria. *Nanomedicine*, 755-764.

Chen, X., Cai, K., Fang, J., Lai, M., Li, J., Hou, Y., Luo, Z., Hu, Y. and Tang, L. (2013). Dual action antibacterial TiO_2 nanotubes incorporated with silver nanoparticles and coated with a quaternary ammonium salt (QAS). *Surface and Coatings Technology*, 158-165.

Cheow, W. S., Chang, M. W. and Hadinoto, K. (2010). Antibacterial efficacy of inhalable antibiotic-encapsulated biodegradable polymeric nanoparticles against *E. coli* biofilm cells. *Journal of biomedical nanotechnology*, 391-403.

Chifiriuc, M. C., Grumezescu, A. M., Andronescu, E., Ficai, A., Cotar, A. I., Grumezescu, V., Bezirtzoglou, E., Lazar, V. and Radulescu, R. (2013). Water dispersible magnetite nanoparticles influence the efficacy of antibiotics against planktonic and biofilm embedded Enterococcus faecalis cells. *Anaerobe*, 14-19.

Costerton, J. (1999). Introduction to biofilm. Intern J Antimicrob Agents 11: 217-221. C osterton JW, Stewart PS, Greenberg EP (1999) Bacterial biofilms: a common cause of persistent infections. *Science*, 1318-1322.

De Freitas, L. M., Calixto, G. M. F., Chorilli, M., Giusti, J. S. M., Bagnato, V. S., Soukos, N. S., Amiji, M. M. and Fontana, C. R. (2016). Polymeric nanoparticle-based photodynamic therapy for chronic periodontitis in vivo. *International journal of molecular sciences*, 769.

De Giglio, E., Trapani, A., Cafagna, D., Ferretti, C., Iatta, R., Cometa, S., Ceci, E., Romanelli, A. and Mattioli-Belmonte, M. (2012). Ciprofloxacin-loaded Chitosan nanoparticles as Titanium Coatings: A Valuable Strategy to Prevent Implant-associated Infections. *Nano Biomedicine & Engineering*.

Dorling, K. M. and Baker, M. J. (2013). Highlighting attenuated total reflection Fourier transform infrared spectroscopy for rapid serum analysis. *Trends in biotechnology*, 327-328.

Du, Y., Luo, X. L., Xu, J. J. and Chen, H. Y. (2007). A simple method to fabricate a chitosan-gold nanoparticles film and its application in glucose biosensor. *Bioelectrochemistry*, 342-347.

Duncan, B., Li, X., Landis, R. F., Kim, S. T., Gupta, A., Wang, L. S., Ramanathan, R., Tang, R., Boerth, J. A. and Rotello, V. M. (2015). Nanoparticle-stabilized capsules for the treatment of bacterial biofilms. *ACS nano*, 7775-7782.

El Badawy, A. M., Silva, R. G., Morris, B., Scheckel, K. G., Suidan, M. T. and Tolaymat, T. M. (2011). Surface charge-dependent toxicity of silver nanoparticles. *Environmental science & technology*, 283-287.

Elizabeth, E., Baranwal, G., Krishnan, A. G., Menon, D. and Nair, M. (2014). ZnO nanoparticle incorporated nanostructured metallic titanium for increased mesenchymal stem cell response and antibacterial activity. *Nanotechnology*, 115101.

Eshed, M., Lellouche, J., Matalon, S., Gedanken, A. and Banin, E. (2012). Sonochemical coatings of ZnO and CuO nanoparticles inhibit Streptococcus mutans biofilm formation on teeth model. *Langmuir*, 12288-12295.

Fisher, L. E., Hook, A. L., Ashraf, W., Yousef, A., Barrett, D. A., Scurr, D. J., Chen, X., Smith, E. F., Fay, M. and Parmenter, C. D. (2015). Biomaterial modification of urinary catheters with antimicrobials to give long-term broadspectrum antibiofilm activity. *Journal of Controlled Release*, 57-64.

Forier, K., Raemdonck, K., De Smedt, S. C., Demeester, J., Coenye, T. and Braeckmans, K. (2014). Lipid and polymer nanoparticles for drug delivery to bacterial biofilms. *Journal of Controlled Release*, 607-623.

Fulaz, S., Vitale, S., Quinn, L. and Casey, E. (2019). Nanoparticle–biofilm interactions: the role of the EPS matrix. *Trends in microbiology*, 915-926.

Garg, A. and Singh, S. (2011). Enhancement in antifungal activity of eugenol in immunosuppressed rats through lipid nanocarriers. *Colloids and Surfaces B: Biointerfaces*, 280-288.

Geilich, B. M. and Webster, T. J. (2013). Reduced adhesion of *Staphylococcus aureus* to ZnO/PVC nanocomposites. 2013 39th Annual Northeast Bioengineering Conference.

Gharbi, A., Humblot, V., Turpin, F., Pradier, C. M., Imbert, C. and Berjeaud, J. M. (2012). Elaboration of antibiofilm surfaces functionalized with antifungal-cyclodextrin inclusion complexes. *FEMS Immunology & Medical Microbiology*, 257-269.

Giri, K., Yepes, L. R., Duncan, B., Parameswaran, P. K., Yan, B., Jiang, Y., Bilska, M., Moyano, D. F., Thompson, M. A. and Rotello, V. M. (2015). Targeting bacterial biofilms via surface engineering of gold nanoparticles. *RSC advances*, 105551-105559.

Grunkemeier, G. L., Jin, R. and Starr, A. (2006). Prosthetic heart valves: objective performance criteria versus randomized clinical trial. *The Annals of thoracic surgery*, 776-780.

Haghighi, F., Roudbar Mohammadi, S., Mohammadi, P., Hosseinkhani, S. and Shipour, R. (2013). Antifungal activity of TiO_2 nanoparticles and EDTA on *Candida albicans* biofilms. *Infection, Epidemiology and Microbiology*, 33-38.

Hajipour, M. J., Fromm, K. M., Ashkarran, A. A., de Aberasturi, D. J., de Larramendi, I. R., Rojo, T., Serpooshan, V., Parak, W. J. and Mahmoudi, M. (2012). Antibacterial properties of nanoparticles. *Trends in biotechnology*, 499-511.

Hazer, D. B., Sakar, M., Dere, Y., Altinkanat, G., Ziyal, M. I. and Hazer, B. (2016). Polymer-based Silver Nanoparticle Coated Pedicle Screws Inhibits Biofilm Formation-a Novel Experimental Model in Rabbits. *Global Spine Journal*, s-0036-1582883-s-1580036-1582883.

Ho, C. H., Odermatt, E. K., Berndt, I. and Tiller, J. C. (2013). Long-term active antimicrobial coatings for surgical sutures based on silver nanoparticles and hyperbranched polylysine. *Journal of Biomaterials Science, Polymer Edition*, 1589-1600.

Horev, B., Klein, M. I., Hwang, G., Li, Y., Kim, D., Koo, H. and Benoit, D. S. (2015). pH-activated nanoparticles for controlled topical delivery of farnesol to disrupt oral biofilm virulence. *ACS nano*, 2390-2404.

Huang, J. F., Zhong, J., Chen, G. P., Lin, Z. T., Deng, Y., Liu, Y. L., Cao, P. Y., Wang, B., Wei, Y. and Wu, T. (2016). A hydrogel-based hybrid theranostic contact lens for fungal keratitis. *ACS nano*, 6464-6473.

Janiszewska, J., Swieton, J., Lipkowski, A. W. and Urbanczyk-Lipkowska, Z. (2003). Low molecular mass peptide dendrimers that express antimicrobial properties. *Bioorganic & medicinal chemistry letters*, 3711-3713.

Jian, H. J., Wu, R. S., Lin, T. Y., Li, Y. J., Lin, H. J., Harroun, S. G., Lai, J. Y. and Huang, C. C. (2017). Super-cationic carbon quantum dots synthesized from spermidine as an eye drop formulation for topical treatment of bacterial keratitis. *ACS nano*, 6703-6716.

Jian, H. J., Yu, J., Li, Y. J., Unnikrishnan, B., Huang, Y. F., Luo, L. J., Ma, D. H. K., Harroun, S. G., Chang, H. T. and Lin, H. J. (2020). Highly adhesive carbon quantum dots from biogenic amines for prevention of biofilm formation. *Chemical Engineering Journal*, 123913.

Johansson, E. M., Crusz, S. A., Kolomiets, E., Buts, L., Kadam, R. U., Cacciarini, M., Bartels, K. M., Diggle, S. P., Cámara, M. and Williams, P. (2008). Inhibition and dispersion of *Pseudomonas aeruginosa* biofilms by glycopeptide dendrimers targeting the fucose-specific lectin LecB. *Chemistry & biology*, 1249-1257.

Jones, N., Ray, B., Ranjit, K. T. and Manna, A. C. (2008). Antibacterial activity of ZnO nanoparticle suspensions on a broad spectrum of microorganisms. *FEMS microbiology letters*, 71-76.

Kaittanis, C., Nath, S. and Perez, J. M. (2008). Rapid nanoparticle-mediated monitoring of bacterial metabolic activity and assessment of antimicrobial susceptibility in blood with magnetic relaxation. *PLoS One*, e3253.

Kassen, R. and Rainey, P. B. (2004). The ecology and genetics of microbial diversity. *Annu. Rev. Microbiol.*, 207-231.

Khan, S., Alam, F., Azam, A. and Khan, A. U. (2012). Gold nanoparticles enhance methylene blue–induced photodynamic therapy: a novel therapeutic approach to inhibit Candida albicans biofilm. *International journal of nanomedicine*, 3245.

Kim, M. H., Yamayoshi, I., Mathew, S., Lin, H., Nayfach, J. and Simon, S. I. (2013). Magnetic nanoparticle targeted hyperthermia of cutaneous *Staphylococcus aureus* infection. *Annals of biomedical engineering*, 598-609.

Kostenko, V., Lyczak, J., Turner, K. and Martinuzzi, R. J. (2010). Impact of silver-containing wound dressings on bacterial biofilm viability and susceptibility to antibiotics during prolonged treatment. *Antimicrobial agents and chemotherapy*, 5120-5131.

Kumar, C. G. and Sujitha, P. (2014). Green synthesis of Kocuran-functionalized silver glyconanoparticles for use as antibiofilm coatings on silicone urethral catheters. *Nanotechnology*, 325101.

Kwon, E. J., Skalak, M., Bertucci, A., Braun, G., Ricci, F., Ruoslahti, E., Sailor, M. J. and Bhatia, S. N. (2017). Porous Silicon Nanoparticle Delivery of Tandem Peptide Anti-Infectives for the Treatment of *Pseudomonas aeruginosa* Lung Infections. *Advanced materials*, 1701527.

Landis, R. F., Gupta, A., Lee, Y. W., Wang, L. S., Golba, B., Couillaud, B., Ridolfo, R., Das, R. and Rotello, V. M. (2017). Cross-linked polymer-stabilized nanocomposites for the treatment of bacterial biofilms. *ACS nano*, 946-952.

Lee, J. H., Kim, Y. G., Cho, M. H. and Lee, J. (2014). ZnO nanoparticles inhibit *Pseudomonas aeruginosa* biofilm formation and virulence factor production. *Microbiological research*, 888-896.

Lee, J. and Ramasamy, M. (2016). Recent nanotechnology approaches for prevention and treatment of biofilm-associated infections on medical devices recent nanotechnology approaches for prevention and treatment of biofilm-associated infections on medical devices. 2016.

Lemire, J. A., Harrison, J. J. and Turner, R. J. (2013). Antimicrobial activity of metals: mechanisms, molecular targets and applications. *Nature Reviews Microbiology*, 371-384.

Li, X., Li, P., Saravanan, R., Basu, A., Mishra, B., Lim, S. H., Su, X., Tambyah, P. A. and Leong, S. S. J. (2014). Antimicrobial functionalization of silicone surfaces with engineered short peptides having broad spectrum antimicrobial and salt-resistant properties. *Acta biomaterialia*, 258-266.

Li, Y., Leung, P., Yao, L., Song, Q. and Newton, E. (2006). Antimicrobial effect of surgical masks coated with nanoparticles. *Journal of Hospital Infection*, 58-63.

Liaqat, I., Bachmann, R.T. and Edyvean, R.G. (2014). Type 2 quorum sensing monitoring, inhibition and biofilm formation in marine microrganisms. *Current microbiology*, 342-351.

Lichter, J. A. and Rubner, M. F. (2009). Polyelectrolyte multilayers with intrinsic antimicrobial functionality: the importance of mobile polycations. *Langmuir*, 7686-7694.

Liu, Y., Busscher, H. J., Zhao, B., Li, Y., Zhang, Z., van der Mei, H. C., Ren, Y. and Shi, L. (2016). Surface-adaptive, antimicrobially loaded, micellar nanocarriers with enhanced penetration and killing efficiency in staphylococcal biofilms. *ACS nano*, 4779-4789.

Liu, Y., Ren, Y., Li, Y., Su, L., Zhang, Y., Huang, F., Liu, J., Liu, J., van Kooten, T. G. and An, Y. (2018). Nanocarriers with conjugated antimicrobials to eradicate pathogenic biofilms evaluated in murine *in vivo* and human *ex vivo* infection models. *Acta biomaterialia*, 331-343.

Ma, H., Darmawan, E. T., Zhang, M., Zhang, L. and Bryers, J. D. (2013). Development of a poly (ether urethane) system for the controlled release of two novel anti-biofilm agents based on gallium or zinc and its efficacy to prevent bacterial biofilm formation. *Journal of Controlled Release*, 1035-1044.

Macherla, C., Sanchez, D. A., Ahmadi, M., Vellozzi, E. M., Friedman, A. J., Nosanchuk, J. D. and Martinez, L. R. (2012). Nitric oxide releasing nanoparticles for treatment of *Candida albicans* burn infections. *Frontiers in microbiology*, 193.

Magalhães, A. P. R., Santos, L. B., Lopes, L. G., Estrela, C. R. d. A., Estrela, C., Torres, É. M., Bakuzis, A. F., Cardoso, P. C. and Carrião, M. S. (2012). Nanosilver application in dental cements. *International Scholarly Research Notices*.

Maity, D., Chandrasekharan, P., Yang, C. T., Chuang, K. H., Shuter, B., Xue, J. M., Ding, J. and Feng, S. S. (2010). Facile synthesis of water-stable magnetite nanoparticles for clinical MRI and magnetic hyperthermia applications. *Nanomedicine*, 1571-1584.

Mirani, Z. A., Khan, M. N., Siddiqui, A., Khan, F., Aziz, M., Naz, S., Ahmed, A. and Khan, S. I. (2018). Ascorbic acid augments colony spreading by reducing biofilm formation of methicillin-resistant *Staphylococcus aureus*. *Iranian journal of basic medical sciences*, 175.

Morones, J. R., Elechiguerra, J. L., Camacho, A., Holt, K., Kouri, J. B., Ramírez, J. T. and Yacaman, M. J. (2005). The bactericidal effect of silver nanoparticles. *Nanotechnology*, 2346.

Mu, H., Tang, J., Liu, Q., Sun, C., Wang, T. and Duan, J. (2016). Potent antibacterial nanoparticles against biofilm and intracellular bacteria. *Scientific reports*, 1-9.

Nadell, C. D., Xavier, J. B. and Foster, K. R. (2008). The sociobiology of biofilms. *FEMS microbiology reviews*, 206-224.

Naik, K., Chatterjee, A., Prakash, H. and Kowshik, M. (2013). Mesoporous TiO2 nanoparticles containing Ag ion with excellent antimicrobial activity at remarkable low silver concentrations. *Journal of biomedical nanotechnology*, 664-673.

Naik, K. and Kowshik, M. (2014). Anti-biofilm efficacy of low temperature processed AgCl–TiO_2 nanocomposite coating. *Materials Science and Engineering: C*, 62-68.

Nava-Ortiz, C. A., Burillo, G., Concheiro, A., Bucio, E., Matthijs, N., Nelis, H., Coenye, T. and Alvarez-Lorenzo, C. (2010). Cyclodextrin-functionalized biomaterials loaded with miconazole prevent *Candida albicans* biofilm formation in vitro. *Acta biomaterialia*, 1398-1404.

Nevius, B. A., Chen, Y. P., Ferry, J. L. and Decho, A. W. (2012). Surface-functionalization effects on uptake of fluorescent polystyrene nanoparticles by model biofilms. *Ecotoxicology*, 2205-2213.

Ohko, Y., Nagao, Y., Okano, K., Sugiura, N., Fukuda, A., Yang, Y., Negishi, N., Takeuchi, M. and Hanada, S. (2009). Prevention of Phormidium tenue biofilm formation by TiO_2 photocatalysis. *Microbes and environments*, 0906010095-0906010095.

Pal, S., Tak, Y. K. and Song, J. M. (2007). Does the antibacterial activity of silver nanoparticles depend on the shape of the nanoparticle? A study of the Gram-negative bacterium *Escherichia coli*. *Applied and environmental microbiology*, 1712-1720.

Park, H., Park, H. J., Kim, J. A., Lee, S. H., Kim, J. H., Yoon, J. and Park, T. H. (2011). Inactivation of *Pseudomonas aeruginosa* PA01 biofilms by hyperthermia using superparamagnetic nanoparticles. *Journal of microbiological methods*, 41-45.

Pelgrift, R. Y. and Friedman, A. J. (2013). Nanotechnology as a therapeutic tool to combat microbial resistance. *Advanced drug delivery reviews*, 1803-1815.

Peng, Z., Ni, J., Zheng, K., Shen, Y., Wang, X., He, G., Jin, S. and Tang, T. (2013). Dual effects and mechanism of TiO_2 nanotube arrays in reducing bacterial colonization and enhancing C3H10T1/2 cell adhesion. *International journal of nanomedicine*, 3093.

Pornpattananangkul, D., Zhang, L., Olson, S., Aryal, S., Obonyo, M., Vecchio, K., Huang, C. M. and Zhang, L. (2011). Bacterial toxin-triggered drug release from gold nanoparticle-stabilized liposomes for the treatment of bacterial infection. *Journal of the american chemical society*, 4132-4139.

Puckett, S. D., Taylor, E., Raimondo, T. and Webster, T. J. (2010). The relationship between the nanostructure of titanium surfaces and bacterial attachment. *Biomaterials*, 706-713.

Qi, L., Xu, Z., Jiang, X., Hu, C. and Zou, X. (2004). Preparation and antibacterial activity of chitosan nanoparticles. *Carbohydrate research*, 2693-2700.

Qu, W., Hooymans, J. M., Qiu, J., de-Bont, N., Gelling, O. J., van der Mei, H. C. and Busscher, H. J. (2013). Nonadhesive, silica nanoparticles based brush coated contact lens cases—Compromising between ease of cleaning and microbial transmission to contact lenses. *Journal of Biomedical Materials Research Part B: Applied Biomaterials*, 640-647.

Ramalingam, K., Frohlich, N. C. and Lee, V. A. (2013). Effect of nanoemulsion on dental unit waterline biofilm. *Journal of Dental Sciences*, 333-336.

Ren, G., Hu, D., Cheng, E. W., Vargas-Reus, M. A., Reip, P. and Allaker, R. P. (2009). Characterisation of copper oxide nanoparticles for antimicrobial applications. *International journal of antimicrobial agents*, 587-590.

Römling, U., Kjelleberg, S., Normark, S., Nyman, L., Uhlin, B. E. and Åkerlund, B. (2014). Microbial biofilm formation: a need to act. *Journal of internal medicine*, 98-110.

Roosjen, A., van der Mei, H. C., Busscher, H. J. and Norde, W. (2004). Microbial adhesion to poly (ethylene oxide) brushes: influence of polymer chain length and temperature. *Langmuir*, 10949-10955.

Sharma, S., Bano, S., Ghosh, A. S., Mandal, M., Kim, H. W., Dey, T. and Kundu, S. C. (2016). Silk fibroin nanoparticles support in vitro sustained antibiotic release and osteogenesis on titanium surface. *Nanomedicine: Nanotechnology, Biology and Medicine*, 1193-1204.

Shi, Z., Neoh, K., Kang, E. and Wang, W. (2006). Antibacterial and mechanical properties of bone cement impregnated with chitosan nanoparticles. *Biomaterials*, 2440-2449.

Shrestha, A. and Kishen, A. (2014). Antibiofilm efficacy of photosensitizer-functionalized bioactive nanoparticles on multispecies biofilm. *Journal of Endodontics*, 1604-1610.

Sinha, R., Karan, R., Sinha, A. and Khare, S. (2011). Interaction and nanotoxic effect of ZnO and Ag nanoparticles on mesophilic and halophilic bacterial cells. *Bioresource technology*, 1516-1520.

Soleimani, N., Mobarez, A. M., Olia, M. S. J. and Atyabi, F. (2015). Synthesis, characterization and effect of the antibacterial activity of chitosan nanoparticles on vancomycin-resistant Enterococcus and other gram negative or gram positive bacteria. *International Journal of Pure and Applied Sciences and Technology*, 14.

Sousa, C., Henriques, M. and Oliveira, R. (2011). Mini-review: antimicrobial central venous catheters–recent advances and strategies. *Biofouling*, 609-620.

Suci, P. A., Berglund, D. L., Liepold, L., Brumfield, S., Pitts, B., Davison, W., Oltrogge, L., Hoyt, K. O., Codd, S. and Stewart, P. S. (2007). High-density targeting of a viral multifunctional nanoplatform to a pathogenic, biofilm-forming bacterium. *Chemistry & biology*, 387-398.

Sun, H., Gao, N., Dong, K., Ren, J. and Qu, X. (2014). Graphene quantum dots-band-aids used for wound disinfection. *ACS nano*, 6202-6210.

Sun, L. M., Zhang, C. L. and Li, P. (2012). Characterization, antibiofilm, and mechanism of action of novel PEG-stabilized lipid nanoparticles loaded with terpinen-4-ol. *Journal of agricultural and food chemistry*, 6150-6156.

Taylor, E. and Webster, T. J. (2011). Reducing infections through nanotechnology and nanoparticles. *International journal of nanomedicine*, 1463.

Thoendel, M., Kavanaugh, J. S., Flack, C. E. and Horswill, A. R. (2011). Peptide signaling in the Staphylococci. *Chemical reviews*, 117-151.

Tuby, R., Gutfreund, S., Perelshtein, I., Mircus, G., Ehrenberg, M., Mimouni, M., Gedanken, A. and Bahar, I. (2016). Fabrication of a Stable and Efficient Antibacterial Nanocoating of Zn CuO on Contact Lenses. *Chem Nano Mat*, 547-551.

Tufail, M.S. and Liaqat, I. (2021). Silver nanoparticles and their applications-A comprehensive review. *Pure and Applied Biology* (PAB), 315-330.

Tufail, M.S., Liaqat, I., Andleeb, S., Naseem, S., Zafar, U., Sadiqa, A., Liaqat, I., Ali, N.M., Bibi, A., Arshad, N. and Saleem, G. (2022). Biogenic Synthesis, Characterization and Antibacterial Properties of Silver Nanoparticles against Human Pathogens. *Journal of Oleo Science*, 21291.

Vasilev, K., Cook, J. and Griesser, H. J. (2009). Antibacterial surfaces for biomedical devices. *Expert review of medical devices*, 553-567.

Walraven, C. J. and Lee, S. A. (2013). Antifungal lock therapy. *Antimicrobial agents and chemotherapy*, 1-8.

Wang, A., Wang, Q., Kudinha, T., Xiao, S. and Zhuo, C. (2016). Effects of fluoroquinolones and azithromycin on biofilm formation of *Stenotrophomonas maltophilia*. *Scientific reports*, 1-11.

Wolcott, R., Rhoads, D., Bennett, M., Wolcott, B., Gogokhia, L., Costerton, J. and Dowd, S. (2010). Chronic wounds and the medical biofilm paradigm. *Journal of wound care*, 45-53.

Wong, M. S., Chen, C. W., Hsieh, C. C., Hung, S. C., Sun, D. S. and Chang, H. H. (2015). Antibacterial property of Ag nanoparticle-impregnated N-doped titania films under visible light. *Scientific reports*, 1-11.

Wu, K., Yang, Y., Zhang, Y., Deng, J. and Lin, C. (2015). Antimicrobial activity and cytocompatibility of silver nanoparticles coated catheters via a biomimetic surface functionalization strategy. *International journal of nanomedicine*, 7241.

Yang, J., Zhang, X., Ma, Y. H., Gao, G., Chen, X., Jia, H. R., Li, Y. H., Chen, Z. and Wu, F. G. (2016). Carbon dot-based platform for simultaneous bacterial distinguishment and antibacterial applications. *ACS applied materials & interfaces*, 32170-32181.

Zhang, F., Zhang, Z., Zhu, X., Kang, E. T. and Neoh, K. G. (2008). Silk-functionalized titanium surfaces for enhancing osteoblast functions and reducing bacterial adhesion. *Biomaterials*, 4751-4759.

Zhang, L., Keogh, S. and Rickard, C. M. (2013). Reducing the risk of infection associated with vascular access devices through nanotechnology: a perspective. *International journal of nanomedicine*, 4453.

Zhang, L., Pornpattananangkul, D., Hu, C. M. and Huang, C. M. (2010). Development of nanoparticles for antimicrobial drug delivery. *Current medicinal chemistry*, 585-594.

Zhang, W., McLamore, E. S., Garland, N., Leon, J. C. and Banks, M. K. (2013). A simple method for quantifying biomass cell and polymer distribution in biofilms. *Journal of microbiological methods*, 367-374.

Zhang, X., Song, J., Klymov, A., Zhang, Y., de Boer, L., Jansen, J. A., van den Beucken, J. J., Yang, F., Zaat, S. A. and Leeuwenburgh, S. C. (2018). Monitoring local delivery of vancomycin from gelatin nanospheres in zebrafish larvae. *International journal of nanomedicine*, 5377.

Zhao, L., Wang, H., Huo, K., Cui, L., Zhang, W., Ni, H., Zhang, Y., Wu, Z. and Chu, P. K. (2011). Antibacterial nano-structured titania coating incorporated with silver nanoparticles. *Biomaterials*, 5706-5716.

Zhao, Y., Dai, X., Wei, X., Yu, Y., Chen, X., Zhang, X. and Li, C. (2018). Near-infrared light-activated thermosensitive liposomes as efficient agents for photothermal and antibiotic synergistic therapy of bacterial biofilm. *ACS applied materials & interfaces*, 14426-14437.

Zharov, V. P., Mercer, K. E., Galitovskaya, E. N. and Smeltzer, M. S. (2006). Photothermal nanotherapeutics and nanodiagnostics for selective killing of bacteria targeted with gold nanoparticles. *Biophysical journal*, 619-627.

Zhu, Y., Ramasamy, M. and Yi, D. K. (2014). Antibacterial activity of ordered gold nanorod arrays. *ACS applied materials & interfaces*, 15078-15085.

Zou, X., Zhang, L., Wang, Z. and Luo, Y. (2016). Mechanisms of the antimicrobial activities of graphene materials. *Journal of the american chemical society*, 2064-2077.

Chapter 8

Chemical Strategies for Combating Microbial Biofilms

Sumaira Mazhar[1,*], Beenish Sarfaras[1], Aisha Waheed Qurashi[1], Sikander Ali[2], Nauman Aftab[2] and Iram Liaqat[3,†]

[1]Department of Biology Lahore Garrison University, Lahore, Pakistan
[2]Institute of Industrial Biotechnology, Government College University, Lahore, Pakistan
[3]Department of Zoology, Government College University, Lahore, Pakistan

Abstract

Microbial biofilms are formed as the result of aggregation of single or diverse microbial species surrounded by a polymeric matrix. Forming a biofilm is a way in which micro-organisms can protect themselves against environmental stresses, including various antibacterial treatments. Biofilm forming microbes are involved in a wide range of diseases and are difficult to control due to the antibiotic resistance of bacteria in a biofilm. Common strategies used for the control of biofilms include inhibition of biofilm formation during the early stages or dispersion of cells forming the biofilm. Although both approaches are extensively used, neither represents a complete and long-term eradication strategy for established biofilms. Consequently, the discovery and development of biofilm eradication agents as comprehensive, stand-alone biofilm treatment options has become a fundamental area of research. A major focus is on various anti-biofilm molecules including

*Corresponding Author's Email: sumairamazhar@lgu.edu.pk.
†Corresponding Author's Email: iramliaq@hotmail.com.

In: Understanding Antibiofilm Activity
Editor: Iram Liaqat
ISBN: 978-1-68507-927-7

herbal active compounds, chelating agents, peptide antibiotics, antibiotics and synthetic chemical compounds along with altering their structures and mechanisms of action. In this chapter, we look at the development of biofilm resistance/tolerance mechanisms and review the chemical strategies used for biofilm remediation.

Keywords: antibiofilm strategies, chelating agents, chemical strategies, polymeric matrix

Introduction

In nature, microbes are present both in a free living state (planktonic) and in the form of aggregations or clumps known as biofilms (Harmsen et al., 2010). In a biofilm, a group of microbes become enclosed in a matrix of self-secreted exopolysaccharides (Prakash et al., 2003; Stoodley et al., 2002). Microbes can form biofilms on almost any surface including living tissue, medical devices industrial systems, laboratories, water and wastewater networks, indeed on any solid surface immersed in, or exposed to, an aqueous solution (Costerton et al., 1999). Almost 90% of microbes have the potential to grow in a biofilm (Li et al., 2017). Biofilm formation offers an excellent defense mechanism by which the microbes protect themselves against a wide range of both biotic and abiotic stresses (Roy et al., 2018). Biofilms consist of about 15% cells, present in the form of microcolonies, and 85% polymeric extracellular substances released by the cells. The chemical composition of this extracellular polymer material varies between species, but generally contains proteins, polysaccharides, and nucleic acids (Flemming et al., 2010). Medically biofilms are very important as they have been observed in the pathogenesis of a number of microbial infections and their removal is difficult (Vestby et al., 2020). One of the biggest challenges to the control of biofilms is the large number of factors influencing the resistance mechanism of the biofilm as represented in Figure 1 (Stewart, 2015).

One of the most important factors affecting biofilm formation and their tolerance mechanisms is the nature of the substratum and surrounding environmental composition and conditions. Surface energy, charge, topography, and stiffness of the substratum material influence microbial adhesion (Song et al., 2015), a primary requirement for colonization (Feng et al., 2015). Once they get attached to the surface, they begin to form aggregates and their attachment act as turning point of their life cycle from the free-living form to the group form (Donlan, 2002). Initially, attached cells are lightly attached with the substratum and it can be easily separated from the surface considered as reversible attachment stage (Palmer et al., 2007). Several appendages formed by microbes like pili, flagella and fimbria are involved in this reversible attachment followed by irreversible attachment assisted by proteins and exopolysaccharides (EPS) (Muhammad et al., 2020; Petrova et al., 2012). Cells that enter into irreversible attachment stage become flat against the surface and difficult to disintegrate using any force. An intracellular signaling molecule, bis-(3′-5′)-cyclic dimeric guanosine monophosphate (c-di-GMP), is produced during the transition from reversible to irreversible attachment. Production of high c-di-GMP in biofilm producing many microbes also regulates the production of EPS production and suppress the flagellar mediated motility in opposing directions (Boyd et al., 2012). cAMP is also reported as intracellular signaling molecule and regulate the biofilm formation through its transition from reversible to irreversible attachment (Ono et al., 2014). Concentrations of the c-di-GMP and cAMP are monitored by diverse changing environmental conditions like carbon and oxygen, and thus the environmental conditions are the reflection of the regulation of surface attachment of biofilm through these second messengers (Toyofuku et al., 2016).

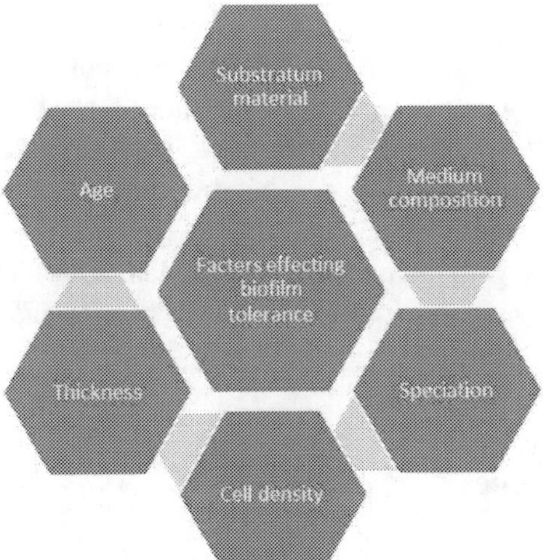

Figure 1. Important factors influencing biofilm tolerance (Stewart, 2015).

Tolerance mechanism of biofilm also varies within with bacterial species. Induction of biofilm specific phenotype is different among different biofilm forming microbial community because of the variation in the expression of active mechanisms to overcome the detrimental effects of antimicrobial agents (Gebreyohannes et al., 2019). Under stress conditions, efflux pumps systems provide tolerance against different chemical agents (Asif et al., 2020; Blanco et al., 2016; Soto, 2013). In the hospitals, multidrug resistant *Acinetobacter baumannii, opportunistic pathogen* is a serious challenge because of its persistence in the environment, and biofilm formation on the wound surface (Gedefie et al., 2021). *Bacillus subtitles and Staphylococcus aureus* cause severe human infections due to their high resistance pattern against widely reported antimicrobial agents because of the activation of efflux pumps (Munita et al., 2016). In *Enterobacter aerogenes, Escherichia coli* and *Klebsiella pneumonia,* the existence of the efflux pump slows down the penetration of hydrophilic solutes which result in the reduction of transmembrane diffusion of lipophilic solutes by down-regulating the 'porin' production (Fernández et al., 2012; Vergalli et al., 2020). Microbes having high density and close proximity also provide them resistance against antibiotics. Increased cell density facilitates the high productions of chemical signals required for cellular communication and social interaction among cells forming biofilm (Li et al., 2012; Shrout et al., 2011). Cellular communication known as quorum sensing in *Vibrio fischeri observed in the form of* bioluminescence occur only at high cell density (Lupp et al., 2005; Verma et al., 2013). In biofilm, microbes help each other through collective actions, division of labor and cooperative activities (Dragoš et al., 2018). Critical cell density is required for biofilm causing infection in host and developing resistance against drugs (Roy et al., 2018). Tolerance potential of microbial biofilm is increased with the increase in age. Mature biofilm is more resistant against antibiotics as compared to the newly formed biofilm (Chen et al., 2020).

Chemical Strategies

Formation of a biofilm contributes significantly to pathogenicity and other environmental problems because of its potential to grow on all available biotic and abiotic surfaces (Roy et al., 2018). [6]. Newly formed biofilms are much easier to treat with antibiotics than older ones (Sharma et al., 2019). Strong communication and interaction through signaling pathways makes older biofilms resistant towards the wide range of antibiotics (Romero et al., 2011). These signaling pathways are mainly quorum sensing (QS) and c-di-GMP signaling. Microbes regulate gene expression using QS mechanisms and mediate signal transduction using c-di-GMP. Microbes have diverse QS regulating genes influencing motility and EPS production (Miller et al., 2001). These genes increase biofilm formation and antibiotic resistance by the reduction of motility and increased production of EPS. Intercellular c-di-GMP content also plays a critical role in the maintenance of biofilms. In high amounts, it reduces motility and increases the production of EPS, while in low amounts, it enhances motility causing dispersal of the biofilm. c-di-GMP level is regulated by a number of proteins in association with different enzymes. In association with diguanylate cyclase, these proteins catalyze its synthesis, while in association with phosphodiesterase these proteins break down c-d-GMA. Microbial species contain many such proteins equipped with diguanylate cyclase and phosphodiesterase activity offering the opportunity for bacteria to quickly regulate production of c-di-GMP in response to changing environmental conditions (Römling et al., 2005). Early biofilm formation and signaling pathways can be targeted in the design of antibiofilm chemical compounds. This chapter highlights the different chemical compounds available for the control of biofilm.

Antimicrobial Peptides

One of the most extensively studied class of antibiofilm compounds as an alternative to antibiotics are antimicrobial peptides (AMPs) (Baltzer et al., 2011). These compounds are naturally produced by plants and animals. Their size and molecular mass varies from 5–90 amino acids and 1- 5 kDa, respectively. Most of them are positively charged known as cationic antimicrobial peptides (Brown et al., 2006), however, negatively charged or anionic forms have also been reported (Harris et al., 2009). Their mechanism of controlling a biofilm is associated with disruption of the cytoplasmic membrane and inhibition enzyme activity or protein folding (Bechinger et al., 2017; Shai, 1999). Antimicrobial peptides have received considerable attention over the last few years as innovative antibiofilm drugs exhibiting broad-spectrum activity. AMPs can eradicate the mature biofilm present nearest the surface (Batoni et al., 2016). Many reports are available on the screening and optimization of AMPs but only few of these have been tested in clinical model systems (Crabbé et al., 2017; de Breij et al., 2018; Lashua et al., 2016).

LL-37 is broad spectrum AMP, derived from human cathelicidins, mostly found in fluid of the body and it is amphipathic in nature (Burton et al., 2009; Nijnik et al., 2009). It decreases attachment of *Pseudomonas aeruginosa* and *S. epidermidis* to plastic surfaces at low concentrations and reduces the thickness of established biofilms at higher concentrations (de la Fuente-Núñez et al., 2015; Hell et al., 2010). A recently reported semi-synthetic peptide, lin-SB056-1, is optimized structurally to enhance its efficacy against microbial biofilms. It displays strong antibacterial activity against *P. aeruginosa* isolated from samples of lung from infected patients (Grassi et al., 2019). AMPs are attractive candidates for the treatment of various human infections caused by multiple drug resistant pathogens. Other reported medically important antimicrobial peptides are listed in Table 1.

Antimicrobial Lipids

Another important class for the control of biofilms are antimicrobial lipids, defined as single-chain lipid amphiphiles mainly consisting of fatty acids and monoglycerides (Yoon et al., 2018). Antimicrobial peptides offer potential candidates as antibiofilm agents through destabilizing the bacterial cell wall. Antimicrobial activity by lipids was observed by Robert Koch in the 1800s for the control of *Bacillus anthracis* (responsible for causing anthrax disease) (Thormar, 2011). Not much attention was paid to this due to the discovery of antibiotics. However, antibiotic resistance in antimicrobial lipids of bacteria is again getting consideration, particular attention is being given to fatty acids and monoglycerides (Desbois et al., 2010).

Through biological assays, both bactericidal and bacteriostatic potential has been observed in antimicrobial lipids, physical factors, the targeted microbe and dose concentration (Desbois and Smith, 2010; Heerklotz, 2008). Although they have been well known for the control of pathogens for some time, they are now being explored as antibiofilm agents. Some studies have shown that glycerol monolaurate has the potential to eradicate biofilm from surfaces (Table 1).

Table 1. Antimicrobial peptides and lipids for the control of biofilm

	Compounds	Targeted Organism	References
Antimicrobial peptides	LL-37	*P. aeruginosa*	(de la Fuente-Núñez and Hancock, 2015)
		S. epidermidis	(Hell et al., 2010)
	Lin-SB056-1	*P. aeruginosa*	(Grassi et al., 2019)
	peptide 1018	*Burkholderia cenocepacia P. aeruginosa, S. aureus, Acinetobacter baumannii, E. coli, Klebsiella pneumoniae, Salmonella typhimurium*	(Reffuveille et al., 2014)
	antibiotics (nisin, subtilin, epidermin and gallidermin)	*S. aureus, Lactococcus lactis, S. epidermidis*	(Parisot et al., 2008; Saising et al., 2012)
	Oritavancin	*S. aureus*	(Allen, 2010; Belley et al., 2009)
	Novispirin G10	*P. aeruginosa*	(Steinstraesser et al., 2002)
Antimicrobial lipids	Glycerol monolaurate	*Listeria monocytogenes*	(Oh et al., 1995)
	Glycerol monolaurate	*S. aureus*	(Schlievert et al., 2012)
	Glycerol monolaurate	*P. aeruginosa*	(Lopes et al., 2019)
	Glycerol monolaurate and aminoglycosides	*S. aureus*	(Hess et al., 2014)
	Glycerol monolaurate	*S. aureus, Enterococcus faecalis*	(Hess et al., 2015)
	Glycerol monolaurate	*S. aureus, Clostridium perfringens, Bacillus subtilis, E. coli*	(Schlievert et al., 2019)

Quaternary Ammonium Compounds

Quaternary ammonium compounds (QACs) represent one of the most effective classes of broad-spectrum bactericidal agents used as disinfectants all over the world (Jiao et al., 2017). The structure of QACs is amphiphilic mainly composed of a hydrophilic quaternary ammonium group hydrophobic alkyl chain. Commonly they are referred as AMP mimics (Zakharova et al., 2019). Exploitation of QACs as disinfectants and antiseptics dates back to the 1930s (Domagk, 1935). They can be prepared easily with a wide variation in structure and versatility when incorporated into consumer products (Jennings et al., 2015), generally work through the disruption of the cell membrane i.e., affecting one of the most fundamental structures in bacteria. Damaging of the cell membrane eventually causes its lysis and leakage of metabolites from the cell leading to bacterial death. Microbial biofilms are thought to be sensitive toward QACs (Druvari et al., 2018; Ioannou et al., 2007). In addition to the killing of bacteria, QACs also prevent their attachment through modification of surface properties (Simoes et al., 2005).

The potential for QACs to control microbes is widely accepted and well-documented. Their antimicrobial effect is regulated by many factors (Ikeda et al., 1986). Their molecular weight has a critical effect on their antimicrobial activity (Chen et al., 2000; Russell, 2003). Artificially polymers impregnated with QACs can be prepared using two methods. First involves quaternization of a reactive precursor polymer generates products with variable degrees of cationization as a result of the unpredictable impact of steric hindrance from neighboring groups. The second method is the copolymerization of monomers having QA functional groups (QA monomers) within a polymer network (Schlievert et al., 2019).

Coating medical devices with antibacterial agents is mostly effective for only a short duration because they leach over time (Ooi et al., 2010). QACs polymers coated on medical devices offer a more attractive method for controlling bacterial infection by inhibiting biofilm formation (Dror et al., 2009).

Chlorine-Releasing Agents

Chlorine-releasing agents (CRAs) exhibit antimicrobial properties through the release of chlorine into the environment which reacts with the required target. CRAs are mostly represented by N-halamines. *N*-halamines can be categorized on the basis of their structure into three groups. These are amine *N*-halamines, amide *N*-halamines and imide *N*-halamines (Worley et al., 1988). Chloramine has recently been synthesized and represents a unique class of *N*-halamines which can be used for biofilm control. 2-amino-4-chloro-6-hydroxy-*s*-triazine (ACHT) is a chloramine-based polymeric material. ACHT can be immobilized onto cotton cellulose which can then discharge chlorine into the targeted area and control the biofilm (Chen, Luo, and Sun, 2007). ACHT shows sporicidal efficacy against spores forming microbes which can be exploited further in biofilm control. The chlorinated fibrous materials effectively inactivated the microbes and could prevent the formation and development of biofilms (Wang et al., 2020). Biofilms formed in drinking water distribution systems (DWDSs) are a potential risk to human health (Flemming et al., 2002). *Acinetobacter, Alcaligenes, Aeromonas, Arthrobacter, Bacillus, Burkholderia, Citrobacter, Enterobacter, Flavobacterium, Klebsiella, Methylobacterium, Moraxella, Pseudomonas, Serratia, Staphylococcus, Sphingomonas, Mycobacterium* and *Xanthomonas* spp. are mostly present in DWDSs (Berry et al., 2006; Simoes et al., 2010) in multi-bacterial biofilms. Penetration of the biofilm by chlorine disintegrates the EPS (De Beer et al., 1994).

One of most commonly and widely used CRA is sodium hypochlorite (NaOCl) (Bloomfield and Arthur, 1989). CRAs release free chlorine which after penetration into the cells bind with cell membrane (Clayton et al., 2021; Lineback et al., 2018). After its binding with the cell membrane, it might cause its disruption which ultimately lead to the bacterial death. It can cause disruption through the inactivation or inhibition of various enzymes or its system. But release of free chlorines into the moist environment in specific amount is not able to inactivate the most tough and resistant species e.g., spores. In dealing with bacterial spores, its cortex and the coat material can be degraded and detached respectively by the usage of the interaction with the hypochlorite ion i.e., OCl^- (Bloomfield et al., 1992). Removal of coat material and degradation of cortex also facilitate the penetration of chlorinated ACHT to their target sites of action or in the protoplast of the spore (Chen, Luo, and Sun, 2007).

Enzymes

Biofilm degradation enzymes i.e., DNase I, Dispersin B (DspB) and a-amylase offer another promising antibiofilm strategy (Khan et al., 2021). These enzymes enhance the efficacy of antibiotics through increased penetration, DNase I, DspB and α-amylase degrade eDNA, biofilm matrix and EPS, respectively (Sun et al., 2013; Tetz et al., 2009). Enzymes not only help in the inhibition of biofilm formation but they also facilitate the removal of mature biofilms (Sharahi et al., 2019). Enzymes involved in the degradation of the microbial matrix are mainly categorized as polysaccharide-degrading enzymes, nucleases or proteases (Kaplan, 2014).

Table 2. Quartnary Ammonium Compounds (QACs) used for the control of biofilm

	Compounds	Targeted Organism	References
Quartnary Ammonium Compounds (QACs)	XF-70 and XF-73	*S. aureus*	(Ooi et al., 2010)
	N-alkylpyridinium bromide	*S. epidermidis,* *E. coli,* *P. aeruginosa*	(Dror et al., 2009)
	1-Alkylpyridinium derivatives, 3-alkyl-1-(2-hydroxyethyl) imidazolium derivatives	*Enterococcus faecalis,* *S. epidermidis,* *S. aureus*	(Hympanova et al., 2020)
	n-alkyl ethyl benzyl, n-alkyl dimethyl benzyl ammonium chloride	*S. aureus, P. aeruginosa, Staphylococcus equorium*	(Shah et al., 2021)
	Gemacide PN 50TM	*Legionella pneumophila*	(Kimiran-Erdem et al., 2007)
	Benzalkonium chloride (BAC), Cetyltrimethyl ammonium bromide (CMAB)	*Pseudomonas fluorescens, Bacillus cereus*	(Araújo et al., 2014)
	Benzalkonium chloride	*Listeria monocytogenes*	(Fazlara et al., 2012)
gChlorine releasing agents	2-amino-4-chloro-6-hydroxy-*s*-triazine (ACHT)	*Bacillus subtilis, P. aeruginosa, E. coli, S. epidermidis*	(Chen et al., 2007)
	Sodium hypochlorite (NaOCl), Sodium dichloroisocyanurate, chloramine-T	*Bacillus subtilis*	(Bloomfield et al., 1989)
	Hypochlorite ions [OCl$^-$], Hypochlorous acid [HOCl]	*E. coli, P. aeruginosa*	(Clayton et al., 2021)
	Sodium hypochlorite (NaOCl),	*S. aureus, P. aeruginosa,*	(Lineback et al., 2018)
	Sodiumdichloroisocyanurate (DCIC)	*P. aeruginosa*	(Rosenthal et al., 1992)
	Sodium hypochlorite (NaOCl)	*Enterococcus faecalis*	(Liu et al., 2010)
	Sodium hypochlorite (NaOCl)	*Klebsiella,* *Pseudomonas,* *Flavobacterium*	(Lin et al., 2017)

Dispersin B released by *Actinobacillus actinomycetemco mitans* is a bacterial glycoside hydrolase having potential to hydrolyzes the poly-*N*-acetylglucosamine (PNAG). In *Escherichia* spp. and *Staphylococcus* spp. PNAG is a core component of the EPS matrix (Ramasubbu et al., 2005). In combination with triclosan, Dispersin B also effectively reduces biofilm formation in *S. aureus*, *S. epidermidis* and *E. coli* (Darouiche et al., 2009; Desrousseaux et al., 2013; Kaplan, 2009). Endolysin produced by bacteriophages is a peptidoglycan hydrolase involved in the disruption of biofilm formation through digestion of the bacterial cell wall (Chan et al., 2015). Another important enzyme contributing to the biofilm inhibition and removal is Proteinase K that functions effectively by the cleavage of matrix proteins (Shukla et al., 2017). Application of trypsin or Proteinase K eradicated *Staphylococcus* biofilms after application of dispersin B (Di Poto et al., 2009). In microbes QS system mediated through acyl-homoserine lactone (AHL), signaling molecule regulate biofilm formation and motility (Huber et al., 2001). Enzymes inhibit the biofilm formation through the degradation of AHL molecule or inhibit it synthesis by causing blockage of synthase proteins (Lade et al., 2014).

Electrolyzed Water

Helicobacter pylori is an example of a bacteria involved in gastrointestinal problems and periodontal disease through oral infection that has become resistant to the major antibiotics traditionally used for its control (Yonezawa et al., 2015). The use of electrolyzed water (EW) has been developed as an oral antiseptic for the control of such oral pathogens (Lee et al., 2006). EW is produced during the electrolysis of a diluted solution of sodium chloride in an electrolysis cell (Qin et al., 2016) and has been used as an aseptic agent for several years. In the food and agricultural area, it has been applied on diverse abiotic surfaces as an antimicrobial agent EW was used as an aseptic agent for several years. In food industry and agricultural area it was applied on diverse abiotic surfaces as s an antimicrobial agent (AL-HAQ et al., 2005). Generation of electrolyzed solutions of superoxidation (ESS) having high antimicrobial potential offer a more stable and safe antiseptic agent for the control of microbes (Lucio-Sauceda et al., 2019). ESS has now become an attractive strategy for tissue asepsis in different areas of medicine such as dentistry, dermatology, surgery and diabetics. In addition to controlling infection, it also has a tendency to enhance tissue regeneration (Medina-Tamayo et al., 2007; Nakae et al., 2000; Piaggesi et al., 2010; Yahagi et al., 2000; You et al., 2017).

Conclusion

The most commonly used chemical method for the inhibition of biofilm formation are antibiotics and biocides. Development of increased resistance in disease causing bacteria against antibiotics is of great concern around the whole world. High-throughput screening of chemical libraries has led to the identification of several small chemical molecules which show promising results in the reduction of biofilms on diverse abiotic surfaces. Antimicrobial peptides, antimicrobial lipids, QACs, CRAs, EW and enzymes are among these chemicals. In addition to the exhibition of antimicrobial activity against wide range of pathogens, these compounds also show reduced development of resistance because of the absence of selective pressures.

References

Al-Haq, M. I., Sugiyama, J. and Isobe, S. (2005). Applications of electrolyzed water in agriculture & food industries. *Food Science and Technology Research*, 135-150.

Allen, N. E. (2010). From vancomycin to oritavancin: the discovery and development of a novel lipoglycopeptide antibiotic. *Anti-Infective Agents in Medicinal Chemistry (Formerly Current Medicinal Chemistry-Anti-Infective Agents)*, 23-47.

Araújo, P. A., Mergulhão, F., Melo, L. and Simões, M. (2014). The ability of an antimicrobial agent to penetrate a biofilm is not correlated with its killing or removal efficiency. *Biofouling*, 675-683.

Asif, L. and Mazhar, S. (2020). 5. A review on the role of genomics, various environmental factors, biochemical and physiological determinants and biofilms responsible for the development of antibiotics resistance in microbes. *Pure and Applied Biology (PAB)*, 2305-2317.

Baltzer, S. A. and Brown, M. H. (2011). Antimicrobial peptides–promising alternatives to conventional antibiotics. *Journal of molecular microbiology and biotechnology*, 228-235.

Batoni, G., Maisetta, G. and Esin, S. (2016). Antimicrobial peptides and their interaction with biofilms of medically relevant bacteria. *Biochimica et Biophysica Acta (BBA)-Biomembranes*, 1044-1060.

Bechinger, B. and Gorr, S.-U. (2017). Antimicrobial peptides: mechanisms of action and resistance. *Journal of dental research*, 254-260.

Belley, A., Neesham-Grenon, E., McKay, G., Arhin, F. F., Harris, R., Beveridge, T., Parr Jr, T. R. and Moeck, G. (2009). Oritavancin kills stationary-phase and biofilm Staphylococcus aureus cells in vitro. *Antimicrobial Agents and Chemotherapy*, 918-925.

Berry, D., Xi, C. and Raskin, L. (2006). Microbial ecology of drinking water distribution systems. *Current opinion in biotechnology*, 297-302.

Blanco, P., Hernando-Amado, S., Reales-Calderon, J. A., Corona, F., Lira, F., Alcalde-Rico, M., Bernardini, A., Sanchez, M. B. and Martinez, J. L. (2016). Bacterial multidrug efflux pumps: much more than antibiotic resistance determinants. *Microorganisms*, 14.

Bloomfield, S. F. and Arthur, M. (1989). Effect of chlorine releasing agents on Bacillus subtilis vegetative cells and spores. *Letters in applied microbiology*, 101-104.

Bloomfield, S. F. and Arthur, M. (1992). Interaction of Bacillus subtilis spores with sodium hypochlorite, sodium dichloroisocyanurate and chloramine-T. *Journal of applied bacteriology*, 166-172.

Boyd, C. D. and O'Toole, G. A. (2012). Second messenger regulation of biofilm formation: breakthroughs in understanding c-di-GMP effector systems. *Annual review of cell and developmental biology*, 439-462.

Brown, K. L. and Hancock, R. E. (2006). Cationic host defense (antimicrobial) peptides. *Current opinion in immunology*, 24-30.

Burton, M. F. and Steel, P. G. (2009). The chemistry and biology of LL-37. *Natural product reports*, 1572-1584.

Chen, C. Z., Beck-Tan, N. C., Dhurjati, P., van Dyk, T. K., LaRossa, R. A. and Cooper, S. L. (2000). Quaternary ammonium functionalized poly (propylene imine) dendrimers as effective antimicrobials: Structure− activity studies. *Biomacromolecules*, 473-480.

Chen, X., Thomsen, T. R., Winkler, H. and Xu, Y. (2020). Influence of biofilm growth age, media, antibiotic concentration and exposure time on *Staphylococcus aureus* and *Pseudomonas aeruginosa* biofilm removal *in vitro*. *BMC microbiology*, 1-11.

Chen, Z., Luo, J. and Sun, Y. (2007). Biocidal efficacy, biofilm-controlling function, and controlled release effect of chloromelamine-based bioresponsive fibrous materials. *Biomaterials*, 1597-1609.

Clayton, G. E., Thorn, R. and Reynolds, D. M. (2021). The efficacy of chlorine-based disinfectants against planktonic and biofilm bacteria for decentralised point-of-use drinking water. *npj Clean Water*, 1-8.

Costerton, J. W., Stewart, P. S. and Greenberg, E. P. (1999). Bacterial biofilms: a common cause of persistent infections. *Science*, 1318-1322.

Crabbé, A., Liu, Y., Matthijs, N., Rigole, P., De La Fuente-Nùñez, C., Davis, R., Ledesma, M. A., Sarker, S., Van Houdt, R. and Hancock, R. E. (2017). Antimicrobial efficacy against Pseudomonas aeruginosa biofilm formation in a three-dimensional lung epithelial model and the influence of fetal bovine serum. *Scientific reports*, 1-13.

Darouiche, R. O., Mansouri, M. D., Gawande, P. V. and Madhyastha, S. (2009). Antimicrobial and antibiofilm efficacy of triclosan and DispersinB® combination. *Journal of Antimicrobial Chemotherapy*, 88-93.

De Beer, D., Srinivasan, R. and Stewart, P. S. (1994). Direct measurement of chlorine penetration into biofilms during disinfection. *Applied and Environmental Microbiology*, 4339-4344.

de Breij, A., Riool, M., Cordfunke, R. A., Malanovic, N., de Boer, L., Koning, R. I., Ravensbergen, E., Franken, M., van der Heijde, T. and Boekema, B. K. (2018). The antimicrobial peptide SAAP-148 combats drug-resistant bacteria and biofilms. *Science translational medicine*.

de la Fuente-Núñez, C. and Hancock, R. E. (2015). Using anti-biofilm peptides to treat antibiotic-resistant bacterial infections. *Postdoc journal: a journal of postdoctoral research and postdoctoral affairs*, 1.

Desbois, A. P. and Smith, V. J. (2010). Antibacterial free fatty acids: activities, mechanisms of action and biotechnological potential. *Applied microbiology and biotechnology*, 1629-1642.

Desrousseaux, C., Sautou, V., Descamps, S. and Traoré, O. (2013). Modification of the surfaces of medical devices to prevent microbial adhesion and biofilm formation. *Journal of hospital Infection*, 87-93.

Di Poto, A., Sbarra, M. S., Provenza, G., Visai, L. and Speziale, P. (2009). The effect of photodynamic treatment combined with antibiotic action or host defence mechanisms on *Staphylococcus aureus* biofilms. *Biomaterials*, 3158-3166.

Domagk, G. (1935). A new class of disinfectants. *Deut. Med. Wochenschr.*, 829-832.

Donlan, R. M. (2002). Biofilms: microbial life on surfaces. *Emerging infectious diseases*, 881.

Dragoš, A., Kiesewalter, H., Martin, M., Hsu, C.-Y., Hartmann, R., Wechsler, T., Eriksen, C., Brix, S., Drescher, K. and Stanley-Wall, N. (2018). Division of labor during biofilm matrix production. *Current Biology*, 1903-1913. e1905.

Dror, N., Mandel, M., Hazan, Z. and Lavie, G. (2009). Advances in microbial biofilm prevention on indwelling medical devices with emphasis on usage of acoustic energy. *Sensors*, 2538-2554.

Druvari, D., Koromilas, N. D., Bekiari, V., Bokias, G. and Kallitsis, J. K. (2018). Polymeric antimicrobial coatings based on quaternary ammonium compounds. *Coatings*, 8.

Fazlara, A. and Ekhtelat, M. (2012). The disinfectant effects of benzalkonium chloride on some important foodborne pathogens. *Am Eurasian J Agric Environ Sci*, 23-29.

Feng, G., Cheng, Y., Wang, S.-Y., Borca-Tasciuc, D. A., Worobo, R. W. and Moraru, C. I. (2015). Bacterial attachment and biofilm formation on surfaces are reduced by small-diameter nanoscale pores: how small is small enough? *npj Biofilms and Microbiomes*, 1-9.

Fernández, L. and Hancock, R. E. (2012). Adaptive and mutational resistance: role of porins and efflux pumps in drug resistance. *Clinical microbiology reviews*, 661-681.

Flemming, H.-C., Percival, S. and Walker, J. (2002). Contamination potential of biofilms in water distribution systems. *Water science and technology: water supply*, 271-280.

Flemming, H. and Wingender, J. (2010). The biofilm matrix. N*at Rev Microbiol.*

Gebreyohannes, G., Nyerere, A., Bii, C. and Sbhatu, D. B. (2019). Challenges of intervention, treatment, and antibiotic resistance of biofilm-forming microorganisms. *Heliyon*, e02192.

Gedefie, A., Demsis, W., Ashagrie, M., Kassa, Y., Tesfaye, M., Tilahun, M., Bisetegn, H. and Sahle, Z. (2021). *Acinetobacter baumannii* biofilm formation and its role in disease pathogenesis: A review. *Infection and Drug Resistance*, 3711.

Grassi, L., Batoni, G., Ostyn, L., Rigole, P., Van den Bossche, S., Rinaldi, A. C., Maisetta, G., Esin, S., Coenye, T. and Crabbé, A. (2019). The antimicrobial peptide lin-SB056-1 and its dendrimeric derivative prevent *Pseudomonas aeruginosa* biofilm formation in physiologically relevant models of chronic infections. *Frontiers in microbiology*, 198.

Harmsen, M., Yang, L., Pamp, S. J. and Tolker-Nielsen, T. (2010). An update on *Pseudomonas aeruginosa* biofilm formation, tolerance, and dispersal. *FEMS Immunology & Medical Microbiology*, 253-268.

Harris, F., Dennison, S. R. and Phoenix, D. A. (2009). Anionic antimicrobial peptides from eukaryotic organisms. *Current Protein and Peptide Science*, 585-606.

Heerklotz, H. (2008). Interactions of surfactants with lipid membranes. *Quarterly reviews of biophysics*, 205-264.

Hell, E., Giske, C., Nelson, A., Römling, U. and Marchini, G. (2010). Human cathelicidin peptide LL37 inhibits both attachment capability and biofilm formation of *Staphylococcus epidermidis*. *Letters in applied microbiology*, 211-215.

Hess, D. J., Henry-Stanley, M. J. and Wells, C. L. (2014). Antibacterial synergy of glycerol monolaurate and aminoglycosides in *Staphylococcus aureus* biofilms. *Antimicrobial Agents and Chemotherapy*, 6970-6973.

Hess, D. J., Henry-Stanley, M. J. and Wells, C. L. (2015). The natural surfactant glycerol monolaurate significantly reduces development of *Staphylococcus aureus* and *Enterococcus faecalis* biofilms. *Surgical infections*, 538-542.

Huber, B., Riedel, K., Hentzer, M., Heydorn, A., Gotschlich, A., Givskov, M., Molin, S. and Eberl, L. (2001). The cep quorum-sensing system of *Burkholderia cepacia* H111 controls biofilm formation and swarming motility. *Microbiology*, 2517-2528.

Hympanova, M., Terlep, S., Markova, A., Prchal, L., Dogsa, I., Pulkrabkova, L., Benkova, M., Marek, J. and Stopar, D. (2020). The Antibacterial Effects of New N-Alkylpyridinium Salts on Planktonic and Biofilm Bacteria [Original Research]. *Frontiers in microbiology*. https://doi.org/10.3389/fmicb.2020.573951.

Ikeda, T., Hirayama, H., Yamaguchi, H., Tazuke, S. and Watanabe, M. (1986). Polycationic biocides with pendant active groups: molecular weight dependence of antibacterial activity. *Antimicrobial Agents and Chemotherapy*, 132-136.

Ioannou, C. J., Hanlon, G. W. and Denyer, S. P. (2007). Action of disinfectant quaternary ammonium compounds against *Staphylococcus aureus*. *Antimicrobial Agents and Chemotherapy*, 296-306.

Jennings, M. C., Minbiole, K. P. and Wuest, W. M. (2015). Quaternary ammonium compounds: an antimicrobial mainstay and platform for innovation to address bacterial resistance. *ACS infectious diseases*, 288-303.

Jiao, Y., Niu, L. N., Ma, S., Li, J., Tay, F. R. and Chen, J.-h. (2017). Quaternary ammonium-based biomedical materials: State-of-the-art, toxicological aspects and antimicrobial resistance. *Progress in polymer science*, 53-90.

K Chan, B. and T Abedon, S. (2015). Bacteriophages and their enzymes in biofilm control. *Current pharmaceutical design*, 85-99.

Kaplan, J. B. (2009). Therapeutic potential of biofilm-dispersing enzymes. *The International journal of artificial organs*, 545-554.

Kaplan, J. B. (2014). Biofilm matrix-degrading enzymes. In *Microbial Biofilms* (pp. 203-213). Springer.

Khan, J., Tarar, S. M., Gul, I., Nawaz, U. and Arshad, M. (2021). Challenges of antibiotic resistance biofilms and potential combating strategies: a review. *3 Biotech*, 1-15.

Kimiran-Erdem, A., Sanli-Yurudu, N. O. and Cotuk, A. (2007). Efficacy of a quaternary ammonium compound against planktonic and sessile populations of different *Legionella pneumophila* strains. *Annals of microbiology*, 121-125.

Lade, H., Paul, D. and Kweon, J. H. (2014). N-acyl homoserine lactone-mediated quorum sensing with special reference to use of quorum quenching bacteria in membrane biofouling control. *BioMed research international*.

Lashua, L. P., Melvin, J. A., Deslouches, B., Pilewski, J. M., Montelaro, R. C. and Bomberger, J. M. (2016). Engineered cationic antimicrobial peptide (eCAP) prevents *Pseudomonas aeruginosa* biofilm growth on airway epithelial cells. *Journal of Antimicrobial Chemotherapy*, 2200-2207.

Lee, S.-H. and Choi, B.-K. (2006). Antibacterial effect of electrolyzed water on oral bacteria. *Journal of Microbiology*, 417-422.

Li, X.-H. and Lee, J.-H. (2017). Antibiofilm agents: A new perspective for antimicrobial strategy. *Journal of Microbiology*, 753-766.

Li, Y.-H. and Tian, X. (2012). Quorum sensing and bacterial social interactions in biofilms. *Sensors*, 2519-2538.

Lin, H., Zhu, X., Wang, Y. and Yu, X. (2017). Effect of sodium hypochlorite on typical biofilms formed in drinking water distribution systems. *Journal of water and health*, 218-227.

Lineback, C. B., Nkemngong, C. A., Wu, S. T., Li, X., Teska, P. J. and Oliver, H. F. (2018). Hydrogen peroxide and sodium hypochlorite disinfectants are more effective against *Staphylococcus aureus* and P*seudomonas aeruginosa* biofilms than quaternary ammonium compounds. *Antimicrobial Resistance & Infection Control*, 1-7.

Liu, H., Wei, X., Ling, J., Wang, W. and Huang, X. (2010). Biofilm formation capability of *Enterococcus faecalis* cells in starvation phase and its susceptibility to sodium hypochlorite. *Journal of endodontics*, 630-635.

Lopes, L. Q. S., de Almeida Vaucher, R., Giongo, J. L., Gündel, A. and Santos, R. C. V. (2019). Characterisation and anti-biofilm activity of glycerol monolaurate nanocapsules against *Pseudomonas aeruginosa*. *Microbial pathogenesis*, 178-185.

Lucio-Sauceda, D. G., Urrutia-Baca, V. H., Gomez-Flores, R., La Garza-Ramos, D., Angélica, M., Tamez-Guerra, P. and Orozco-Flores, A. (2019). Antimicrobial and Anti-Biofilm Effect of an Electrolyzed Superoxidized Solution at Neutral-pH against *Helicobacter pylori*. *BioMed research international*.

Lupp, C. and Ruby, E. G. (2005). Vibrio fischeri uses two quorum-sensing systems for the regulation of early and late colonization factors. *Journal of bacteriology*, 3620-3629.

Medina-Tamayo, J., Sánchez-Miranda, E., Balleza-Tapia, H., Ambriz, X., Cid, M., González-Espinosa, D., Gutiérrez, A. and González-Espinosa, C. (2007). Superoxidized solution inhibits IgE-antigen-induced degranulation and cytokine release in mast cells. *International immunopharmacology*, 1013-1024.

Miller, M. B. and Bassler, B. L. (2001). Quorum sensing in bacteria. *Annual Reviews in Microbiology*, 165-199.

Muhammad, M. H., Idris, A. L., Fan, X., Guo, Y., Yu, Y., Jin, X., Qiu, J., Guan, X. and Huang, T. (2020). Beyond risk: bacterial biofilms and their regulating approaches. *Frontiers in microbiology*, 928.

Munita, J. M. and Arias, C. A. (2016). Mechanisms of antibiotic resistance. *Microbiology spectrum*, 4.2. 15.

Nakae, H. and Inaba, H. (2000). Effectiveness of electrolyzed oxidized water irrigation in a burn-wound infection model. *Journal of Trauma and Acute Care Surgery*, 511-514.

Nijnik, A. and Hancock, R. E. (2009). The roles of cathelicidin LL-37 in immune defences and novel clinical applications. *Current opinion in hematology*, 41-47.

Oh, D.-H. and Marshall, D. L. (1995). Destruction of *Listeria monocytogenes* biofilms on stainless steel using monolaurin and heat. *Journal of food protection*, 251-255.

Ono, K., Oka, R., Toyofuku, M., Sakaguchi, A., Hamada, M., Yoshida, S. and Nomura, N. (2014). cAMP signaling affects irreversible attachment during biofilm formation by *Pseudomonas aeruginosa* PAO1. *Microbes and environments*, 104-106.

Ooi, N., Miller, K., Randall, C., Rhys-Williams, W., Love, W. and Chopra, I. (2010). XF-70 and XF-73, novel antibacterial agents active against slow-growing and non-dividing cultures of *Staphylococcus aureus* including biofilms. *Journal of Antimicrobial Chemotherapy*, 72-78.

Palmer, J., Flint, S. and Brooks, J. (2007). Bacterial cell attachment, the beginning of a biofilm. *Journal of Industrial Microbiology and Biotechnology*, 577-588.

Parisot, J., Carey, S., Breukink, E., Chan, W. C., Narbad, A. and Bonev, B. (2008). Molecular mechanism of target recognition by subtilin, a class I lanthionine antibiotic. *Antimicrobial Agents and Chemotherapy*, 612-618.

Petrova, O. E. and Sauer, K. (2012). Sticky situations: key components that control bacterial surface attachment. *Journal of bacteriology*, 2413-2425.

Piaggesi, A., Goretti, C., Mazzurco, S., Tascini, C., Leonildi, A., Rizzo, L., Tedeschi, A., Gemignani, G., Menichetti, F. and Del Prato, S. (2010). A randomized controlled trial to examine the efficacy and safety of a new super-oxidized solution for the management of wide postsurgical lesions of the diabetic foot. *The international journal of lower extremity wounds*, 10-15.

Prakash, B., Veeregowda, B. and Krishnappa, G. (2003). Biofilms: a survival strategy of bacteria. *Current science*, 1299-1307.

Qin, M., Molitor, H., Brazil, B., Novak, J. T. and He, Z. (2016). Recovery of nitrogen and water from landfill leachate by a microbial electrolysis cell–forward osmosis system. *Bioresource technology*, 485-492.

Ramasubbu, N., Thomas, L., Ragunath, C. and Kaplan, J. (2005). Structural analysis of dispersin B, a biofilm-releasing glycoside hydrolase from the periodontopathogen *Actinobacillus actinomycetemcomitans*. *Journal of molecular biology*, 475-486.

Reffuveille, F., Haney, E., Straus, S. and Hancock, R. (2014). Broad-spectrum anti-biofilm peptide that targets a cellular stress responses. *PLoS Pathog*, e1004152.

Romero, D., Traxler, M. F., López, D. and Kolter, R. (2011). Antibiotics as signal molecules. *Chemical reviews*, 5492-5505.

Römling, U., Gomelsky, M. and Galperin, M. Y. (2005). C-di-GMP: the dawning of a novel bacterial signalling system. *Molecular microbiology*, 629-639.

Rosenthal, R. A., Schlitzen, R. L., McNamee, L. S., Dassanayake, N. L. and Amass, R. (1992). Antimicrobial activity of organic chlorine releasing compounds. *Journal of the British Contact Lens Association*, 81-84.

Roy, R., Tiwari, M., Donelli, G. and Tiwari, V. (2018). Strategies for combating bacterial biofilms: A focus on anti-biofilm agents and their mechanisms of action. *Virulence*, 522-554.

Russell, A. (2003). Biocide use and antibiotic resistance: the relevance of laboratory findings to clinical and environmental situations. *The Lancet infectious diseases*, 794-803.

Saising, J., Dube, L., Ziebandt, A.-K., Voravuthikunchai, S. P., Nega, M. and Götz, F. (2012). Activity of gallidermin on Staphylococcus aureus and Staphylococcus epidermidis biofilms. *Antimicrobial Agents and Chemotherapy*, 5804-5810.

Schlievert, P. M., Kilgore, S. H., Seo, K. S. and Leung, D. Y. (2019). Glycerol monolaurate contributes to the antimicrobial and anti-inflammatory activity of human milk. *Scientific reports*, 1-9.

Schlievert, P. M. and Peterson, M. L. (2012). Glycerol monolaurate antibacterial activity in broth and biofilm cultures. *PloS one*, e40350.

Shah, K. and Muriana, P. M. (2021). Efficacy of a Next Generation Quaternary Ammonium Chloride Sanitizer on Staphylococcus and Pseudomonas Biofilms and Practical Application in a Food Processing Environment. *Applied Microbiology*, 89-103.

Shai, Y. (1999). Mechanism of the binding, insertion and destabilization of phospholipid bilayer membranes by α-helical antimicrobial and cell non-selective membrane-lytic peptides. *Biochimica et Biophysica Acta (BBA)-Biomembranes*, 55-70.

Sharahi, J. Y., Azimi, T., Shariati, A., Safari, H., Tehrani, M. K. and Hashemi, A. (2019). Advanced strategies for combating bacterial biofilms. *Journal of cellular physiology*, 14689-14708.

Sharma, D., Misba, L. and Khan, A. U. (2019). Antibiotics versus biofilm: an emerging battleground in microbial communities. *Antimicrobial Resistance & Infection Control*, 1-10.

Shrout, J. D., Tolker-Nielsen, T., Givskov, M. and Parsek, M. R. (2011). The contribution of cell-cell signaling and motility to bacterial biofilm formation. *MRS bulletin*, 367-373.

Shukla, S. K. and Rao, T. S. (2017). Staphylococcus aureus biofilm removal by targeting biofilm-associated extracellular proteins. *The Indian journal of medical research*, S1.

Simoes, L. C., Simoes, M. and Vieira, M. J. (2010). Influence of the diversity of bacterial isolates from drinking water on resistance of biofilms to disinfection. *Applied and Environmental Microbiology*, 6673-6679.

Simoes, M., Pereira, M. O. and Vieira, M. J. (2005). Effect of mechanical stress on biofilms challenged by different chemicals. *Water Research*, 5142-5152.

Song, F., Koo, H. and Ren, D. (2015). Effects of material properties on bacterial adhesion and biofilm formation. *Journal of dental research*, 1027-1034.

Soto, S. M. (2013). Role of efflux pumps in the antibiotic resistance of bacteria embedded in a biofilm. *Virulence*, 223-229.

Steinstraesser, L., Tack, B. F., Waring, A. J., Hong, T., Boo, L. M., Fan, M.-H., Remick, D. I., Su, G. L., Lehrer, R. I. and Wang, S. C. (2002). Activity of novispirin G10 against Pseudomonas aeruginosa in vitro and in infected burns. *Antimicrobial Agents and Chemotherapy*, 1837-1844.

Stewart, P. S. (2015). Antimicrobial tolerance in biofilms. *Microbiology spectrum*, 3.3. 07.

Stoodley, P., Sauer, K., Davies, D. G. and Costerton, J. W. (2002). Biofilms as complex differentiated communities. *Annual Reviews in Microbiology*, 187-209.

Sun, F., Qu, F., Ling, Y., Mao, P., Xia, P., Chen, H. and Zhou, D. (2013). Biofilm-associated infections: antibiotic resistance and novel therapeutic strategies. *Future Microbiology*, 877-886.

Tetz, G. V., Artemenko, N. K. and Tetz, V. V. (2009). Effect of DNase and antibiotics on biofilm characteristics. *Antimicrobial Agents and Chemotherapy*, 1204-1209.

Thormar, H. (2011). Antibacterial effects of lipids: historical review (1881 to 1960). *Lipids and Essential Oils*, 25.

Toyofuku, M., Inaba, T., Kiyokawa, T., Obana, N., Yawata, Y. and Nomura, N. (2016). Environmental factors that shape biofilm formation. *Bioscience, biotechnology, and biochemistry*, 7-12.

Vergalli, J., Bodrenko, I. V., Masi, M., Moynié, L., Acosta-Gutiérrez, S., Naismith, J. H., Davin-Regli, A., Ceccarelli, M., Van den Berg, B. and Winterhalter, M. (2020). Porins and small-molecule translocation across the outer membrane of Gram-negative bacteria. *Nature Reviews Microbiology*, 164-176.

Verma, S. C. and Miyashiro, T. (2013). Quorum sensing in the squid-Vibrio symbiosis. *International journal of molecular sciences*, 16386-16401.

Vestby, L. K., Grønseth, T., Simm, R. and Nesse, L. L. (2020). Bacterial biofilm and its role in the pathogenesis of disease. *Antibiotics*, 59.

Wang, F., Huang, L., Zhang, P., Si, Y., Yu, J. and Ding, B. (2020). Antibacterial N-halamine fibrous materials. *Composites Communications*, 100487.

Worley, S. D., Williams, D. and Crawford, R. A. (1988). Halamine water disinfectants. *Critical Reviews in Environmental Science and Technology*, 133-175.

Yahagi, N., Kono, M., Kitahara, M., Ohmura, A., Sumita, O., Hashimoto, T., Hori, K., Ning-Juan, C., Woodson, P. and Kubota, S. (2000). Effect of electrolyzed water on wound healing. *Artificial Organs*, 984-987.

Yonezawa, H., Osaki, T. and Kamiya, S. (2015). Biofilm formation by Helicobacter pylori and its involvement for antibiotic resistance. *BioMed research international*.

Yoon, B. K., Jackman, J. A., Valle-González, E. R. and Cho, N.-J. (2018). Antibacterial free fatty acids and monoglycerides: biological activities, experimental testing, and therapeutic applications. *International journal of molecular sciences*, 1114.

You, H. S., Fadriquela, A., Sajo, M. E. J., Bajgai, J., Ara, J., Kim, C. S., Kim, S.-K., Oh, J. R., Shim, K. Y. and Lim, H. K. (2017). Wound healing effect of slightly acidic electrolyzed water on cutaneous wounds in hairless mice via immune-redox modulation. *Biological and Pharmaceutical Bulletin*, 1423-1431.

Zakharova, L. Y., Pashirova, T. N., Doktorovova, S., Fernandes, A. R., Sanchez-Lopez, E., Silva, A. M., Souto, S. B. and Souto, E. B. (2019). Cationic surfactants: Self-assembly, structure-activity correlation and their biological applications. *International journal of molecular sciences*, 5534.

About The Editor

Iram Liaqat, Ph.D.

Associate Professor
Department of Zoology
Government College University
Lahore, Pakistan
iramliaq@hotmail.com

List of Contributors

Gustavo Ramos Salles Ferreira
Thiago Henrique Napoleão
Nazish Mazhar Ali
Bushra Mazhar
Maza Rehman
Sikander Ali
Iram Liaqat
Aisha Waheed Qurashi
Sumaira Mazhar
Beenish Sarfaraz
Nauman Aftab
Beenish Sarfraz
Aisha Sultan
Noor Muhammad
Waiza Ansar
Faiza Bashir
Beenish Sarfaras

Index

A

B

C

D

G

H

J

K

L

M

N

O

P

Q

R

S

T

U